Lecture Notes in Mathematics 2247

More information about this series at http://www.springer.com/series/304

Holger Kammeyer

Introduction to ℓ^2-invariants

Holger Kammeyer
Institute for Algebra and Geometry
Karlsruhe Institute of Technology
Karlsruhe, Germany

ISSN 0075-8434 ISSN 1617-9692 (electronic)
Lecture Notes in Mathematics
ISBN 978-3-030-28296-7 ISBN 978-3-030-28297-4 (eBook)
https://doi.org/10.1007/978-3-030-28297-4

Mathematics Subject Classification (2010): Primary: 55N25, 57Q10; Secondary: 46L10, 57M10

This Springer imprint is published by the registered company Springer Nature Switzerland AG.
The registered company address is: Gewerbestrasse 11, 6330 Cham, Switzerland

Preface

A fundamental idea in algebraic topology is to study a space via an associated chain complex. If the space carries a CW structure, the cellular chain groups capture the number of cells in each degree and the boundary maps reflect how the cells are glued to the previous skeleton. In this sense, the chain complex extracts and bundles what is topologically relevant. Classical invariants, like Betti numbers and Reidemeister torsion, emerge from the chain complex by taking dimensions and determinants. One can repeat this process for the finite coverings of the space. But infinite coverings have infinitely generated chain groups even if the base space is compact. So dimensions might be infinite and determinants are not even defined.

ℓ^2-Invariants cope with this infinite setting. One observes that the deck transformation group G acts on the covering space, hence on the chain groups, and turns them into finitely generated modules over the group ring $\mathbb{Z}G$. But depending on G, this ring might be large (neither left- nor right-Noetherian) and accordingly, the category of $\mathbb{Z}G$-modules has no useful notion of rank or dimension, let alone determinant. So algebraically, and without further assumption on G, we are stuck. But salvation comes from functional analysis: the functor ℓ^2-*completion* turns finitely generated modules over $\mathbb{Z}G$ into finitely generated *Hilbert modules* over the *group von Neumann algebra* $\mathcal{L}(G)$. This category is decisively better behaved: it comes endowed with equivariant versions of all the basic notions of linear algebra: trace, dimension, and determinant. Correspondingly, the ℓ^2-completed chain complex yields ℓ^2-*Betti numbers* and ℓ^2-*torsion*, the ℓ^2-counterparts of Betti numbers, and Reidemeister torsion. These will be the protagonists of this text.

As the reader might have noticed, already defining ℓ^2-Betti numbers and ℓ^2-torsion comes at a price. Sound knowledge both in algebraic topology and functional analysis is required from any student who seriously wants to work with these objects. In an attempt to lower the high entry level to the field, we decided to assume no prior exposure to functional analysis whatsoever, and the course will actually start with the definition of Hilbert space. In contrast, the reader should be familiar with the basic concepts of algebraic topology: fundamental group, covering theory, (co-)homology, CW complexes, and the elementary notions of category theory.

As such, the text at hand is designed for graduate students after a first course on algebraic topology. It has grown out of a lecture given at Karlsruhe, a mini course at the Borel Seminar in Les Diablerets, and some introductory talks the author has given on different occasions. Since ℓ^2-invariants have popped up in contexts as diverse as differential geometry, geometric group theory, 3-manifolds, operator algebras, ergodic theory, cohomology of arithmetic groups, and even Turing machines and quantum groups, it is hoped that also the researcher from another field will find these notes useful for introducing herself to these surprisingly powerful tools.

A rough overview of the contents of this text is presented in the subsequent introductory chapter. Let us only say here that the text ends with surveying a couple of recent research developments in which ℓ^2-invariants have played a major role. In this sense, we hope that the course provides a little more than merely a quick introduction to the field. It is however not meant to be a sequel to Lück's treatment [117]. Instead, we intended to write a shorter account, giving more extensive explanations in the foundational chapters and sparing technical details in the more advanced sections. While several new developments since 2002 were taken up, many more have been left out. It would most definitely be time for a new systematic record.

It is my pleasure to thank Jakob Albers, Sabine Braun, Dawid Kielak, Benjamin Waßermann, and the anonymous referees for many helpful comments and suggestions on previous versions of these notes. In addition, I acknowledge funding from the Deutsche Forschungsgemeinschaft within the priority program "Geometry at Infinity" and I am grateful for the hospitality of the Hausdorff Research Institute for Mathematics in Bonn during the junior trimester program "Topology" from which the text has benefited in various ways.

Karlsruhe, Germany   Holger Kammeyer
June 2019

Contents

Chapter 1
Introduction

A reasonable space, say a connected CW complex X, often does not come alone. It brings along the family of Galois coverings $\{\overline{X}_N\}_{N \trianglelefteq G}$ for $G = \pi_1(X)$. The spaces $\overline{X}_N$ come equipped with a nice (free, cellular) action of the group G/N by deck transformations. This is one of many reasons why modern topology seeks to recover classical achievements in an equivariant setting. Let us consider an easy example, the Betti numbers $b_n(X) = \dim_{\mathbb{C}} H_n(X; \mathbb{C})$ for compact X; and let us concentrate on the most important covering: the universal one $\widetilde{X} = \overline{X}_{\{e\}}$.

For every n-cell in X we fix one of the G-many lifts to an n-cell in $\widetilde{X}$. These choices yield a description of the cellular chain complex $C_*(\widetilde{X}; \mathbb{C})$ as

$$\cdots \longrightarrow (\mathbb{C}G)^{k_{n+1}} \longrightarrow (\mathbb{C}G)^{k_n} \longrightarrow (\mathbb{C}G)^{k_{n-1}} \longrightarrow \cdots$$

where k_n is the number of n-cells. Here $\mathbb{C}G$ is just the free $\mathbb{C}$-vector space with basis G and the unit element in G corresponds to the chosen cell. Recall that the chain modules $C_n(\widetilde{X}; \mathbb{C}) = H_n(\widetilde{X}_n, \widetilde{X}_{n-1}; \mathbb{C})$ are defined by the singular homology of the n-skeleton relative to the $(n-1)$-skeleton. Therefore the G-action on $\widetilde{X}$ by cellular homeomorphisms induces a G-action on $C_*(\widetilde{X}; \mathbb{C})$. In the above picture this action is just given by translating the basis vectors. The differentials are G-equivariant by naturality.

One idea to come up with equivariant Betti numbers would be to find some kind of equivariant dimension "$\dim_{\mathbb{C}G}$", defined on G-invariant subquotients of some $(\mathbb{C}G)^k$, and set "$b_n^G(\widetilde{X}) = \dim_{\mathbb{C}G} H_n(\widetilde{X}; \mathbb{C})$". Of course, any decent such "$\dim_{\mathbb{C}G}$" must take nonnegative values and satisfy the two relations $\dim_{\mathbb{C}G} \mathbb{C}G = 1$ and $\dim_{\mathbb{C}G} V = \dim_{\mathbb{C}G} U + \dim_{\mathbb{C}G} W$ for a short exact sequence

$$0 \longrightarrow U \longrightarrow V \longrightarrow W \longrightarrow 0.$$

But this is where the trouble starts. Say $X = S^1 \vee S^1$ so that $G = F_2$ is the free group on two letters. Since X has one 0-cell and two 1-cells, the chain complex

© Springer Nature Switzerland AG 2019
H. Kammeyer, *Introduction to ℓ^2-invariants*, Lecture Notes in Mathematics 2247,
https://doi.org/10.1007/978-3-030-28297-4_1

$C_*(\widetilde{X}; \mathbb{C})$ is of the form

$$0 \longrightarrow (\mathbb{C}F_2)^2 \xrightarrow{d_1} \mathbb{C}F_2 \longrightarrow 0.$$

Now recall that $\widetilde{X}$ is a tree. As a consequence, every nonzero finite linear combination of 1-cells in $C_1(\widetilde{X}; \mathbb{C}) \cong (\mathbb{C}F_2)^2$ must have an edge in its support without neighbor on one end. But d_1 sends an edge to the difference of its end points; so the lonely end survives and d_1 is injective! Hence the sequence

$$0 \longrightarrow (\mathbb{C}F_2)^2 \xrightarrow{d_1} \mathbb{C}F_2 \longrightarrow \operatorname{coker} d_1 \longrightarrow 0$$

is short exact and buries all hopes to find "$\dim_{\mathbb{C}G}$" as desired. Having said that, here is a glimpse of light that should help us all recover from the shock. Say not only finite linear combinations of edges were allowed but also infinite ones, as long as these are only *square-summable*. Then Fig. 1.1 shows an element $x \in \ker d_1$.

The central node in this picture is as good as any other: we can shift $x \mapsto gx$ for any $g \in F_2$ which illustrates that $\ker d_1$ is an F_2-invariant subspace of $(\ell^2 F_2)^2$.

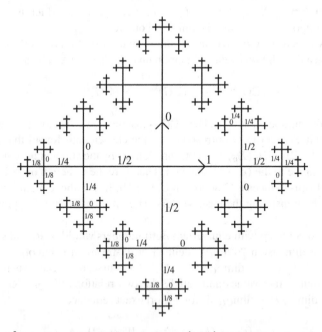

Fig. 1.1 An ℓ^2-1-cycle in the universal covering of $S^1 \vee S^1$. The edge on the right hand side of the center point has coefficient 1. From there on, two of any three neighboring edges obtain half the previous coefficient and the remaining edge gets the coefficient 0. The series formed by the squares of the coefficients converges to 3

Here $\ell^2 F_2$ is the case $G = F_2$ of the general definition

$$\ell^2 G = \left\{ \textstyle\sum_{g \in G} c_g g : \sum_{g \in G} |c_g|^2 < \infty \right\}.$$

The condition that the formal sums in $\ell^2 G$ have square summable (complex) coefficients is also known as the ℓ^2-*condition*. It effects that $\ell^2 G$ has a natural inner product which turns it into a *complete* normed space. Said differently, $\ell^2 G$ is a *Hilbert space*. We remark that for infinite G, the normed space $\mathbb{C}G = \bigoplus_G \mathbb{C}$ is not complete while $\mathbb{C}^G = \prod_G \mathbb{C}$ is not even normable.

The discussion thus far suggests that we should be dealing with closed G-invariant subspaces of $(\ell^2 G)^k$. These are known as *Hilbert G-modules*. It turns out that for Hilbert modules, dimension with the postulated properties can be defined. This so called *von Neumann dimension* "$\dim_{\mathcal{R}(G)}$" can take any nonnegative real number as value. It paves the way for the definition

$$b_n^{(2)}(\widetilde{X}) = \dim_{\mathcal{R}(G)} H_n^{(2)}(\widetilde{X})$$

of ℓ^2-*Betti numbers*, our first and foremost example of an ℓ^2-*invariant*.

If one overcomes a good deal of technical difficulties only to define a variation of a well-known invariant, then it is fair to raise an eyebrow and ask "What's it all good for?". Well, a good way to corroborate the usefulness of a new method is to show that it answers seemingly unrelated questions. Here is an example.

Conjecture 1.1 (Kaplansky) Let G be a torsion-free group. Then the group ring $\mathbb{Q}G$ has no nontrivial zero divisors.

The *group ring* $\mathbb{Q}G$ is the $\mathbb{Q}$-vector space with basis G and linear multiplication defined on the basis by composition in the group. Kaplansky is asking if $a \cdot b = 0$ in $\mathbb{Q}G$ implies $a = 0$ or $b = 0$, an entirely algebraic question.

Theorem 1.2 *Let G be torsion-free. Suppose $b_n^{(2)}(\overline{X}) \in \mathbb{Z}_{\geq 0}$ for $n \geq 0$ whenever $\overline{X}$ is a Galois covering of a connected, compact CW complex X whose deck transformation group embeds into G. Then the Kaplansky conjecture holds true for G.*

We got used to algebra answering questions in topology. This theorem is an instance of the reverse phenomenon. A particular case of the so called *Atiyah conjecture* says that the hypothesis of Theorem 1.2 should always be valid. We will discuss the background on this conjecture, report on recent progress, and see that it is now known for a fairly good deal of groups.

Let us look at a second example. A *closed hyperbolic manifold* is a compact quotient $\mathbb{H}^n / \Gamma$ of hyperbolic n-space $\mathbb{H}^n$ by a torsion-free discrete subgroup $\Gamma \subset \mathrm{Isom}(\mathbb{H}^n)$.

Theorem 1.3 *A closed hyperbolic manifold does not permit any nontrivial action by the circle group.*

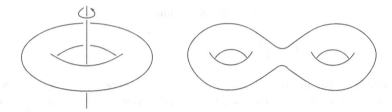

Fig. 1.2 The left hand manifold admits no hyperbolic structure. The right hand manifold admits no nontrivial action by the circle group

To be fair, we should say that this theorem was known long before the advent of ℓ^2-invariants. But ℓ^2-invariants give a particularly clean line of reasoning: Both ℓ^2-Betti numbers and yet to be defined ℓ^2-*torsion* obstruct nontrivial S^1-actions. Even-dimensional hyperbolic manifolds have a nonzero middle ℓ^2-Betti number while odd-dimensional ones have nonzero ℓ^2-torsion. Figure 1.2 is a helpful reminder of the situation.

A hyperbolic manifold is an example of an *aspherical* space: the universal covering is contractible. This leads us to yet another outcome of ℓ^2-invariants.

Conjecture 1.4 (Hopf) Let M be a $2n$-dimensional closed aspherical manifold. Then $(-1)^n \chi(M) \geq 0$.

In the original statement, Hopf discussed the sign of the Euler characteristic $\chi(M)$ in terms of curvature. The above formula was his prediction for non-positively curved manifolds which are aspherical by the so called *Hadamard theorem*. Similar to the classical case, we have an Euler-Poincaré formula $\chi(M) = \sum_{n\geq 0}(-1)^n b_n^{(2)}(\widetilde{M})$ expressing the Euler characteristic in terms of ℓ^2-Betti numbers. This is why the Hopf conjecture is a consequence of the following conjecture.

Conjecture 1.5 (Singer) Let M be an m-dimensional closed aspherical manifold with $b_n^{(2)}(\widetilde{M}) > 0$. Then $2n = m$.

Indeed, ℓ^2-Betti numbers are nonnegative by definition, so for a closed aspherical $2n$-manifold M, the Singer conjecture implies

$$(-1)^n \chi(M) = (-1)^n(-1)^n b_n^{(2)}(\widetilde{M}) = b_n^{(2)}(\widetilde{M}) \geq 0.$$

Now that we elaborated on the interest in studying ℓ^2-Betti numbers $b_n^{(2)}(\widetilde{X})$, let us ponder how they are related to the ordinary Betti numbers $b_n(X)$. As just said, we have $\sum_n(-1)^n b_n^{(2)}(\widetilde{X}) = \sum_n(-1)^n b_n(X)$. But already for the k-torus $\mathbb{T}^k$, the existence of circle actions implies $b_n^{(2)}(\widetilde{\mathbb{T}^k}) = 0$ for all $n \geq 0$ which drastically contrasts with the classical Betti numbers $b_n(\mathbb{T}^k) = \binom{k}{n}$. So the individual Betti number $b_n(X)$ cannot be related to $b_n^{(2)}(\widetilde{X})$ in any all too apparent way. This is maybe not so surprising as $b_n^{(2)}(\widetilde{X})$ is defined in terms of the deck transformation action $G \curvearrowright \widetilde{X}$ on the universal covering whereas the classical Betti number $b_n(X)$ is computed "downstairs" with no dependency on coverings whatsoever.

For a *finite* d-sheeted Galois covering $\overline{X} \to X$, it is however easy to see that $b_n^{(2)}(\overline{X}) = b_n(\overline{X})/d$. So one could hope that for larger and larger finite coverings, the number $b_n(\overline{X})/d$ should give a better and better approximation of $b_n^{(2)}(\widetilde{X})$. More precisely, let us consider sequences $b_n(\overline{X_i})/[G : G_i]$ for towers of finite Galois coverings $\cdots \to \overline{X}_2 \to \overline{X}_1 \to X$ associated with nested chains $G_1 \geq G_2 \geq \cdots$ of finite index normal subgroups $G_i \trianglelefteq G$. The hope would be to obtain $b_n^{(2)}(\widetilde{X})$ in the limit "$\overline{X}_i \to \widetilde{X}$", or correspondingly "$G_i \to \{1\}$", which we can express mathematically as $\bigcap_i G_i = \{1\}$. Such *residual chains* of finite index normal subgroups in G with trivial total intersection may or may not exist. If they do exist, then G is called *residually finite*. Many groups occurring in practice are residually finite, including finitely generated linear groups and fundamental groups of 3-manifolds. *Lück's approximation theorem* asserts the desired asymptotic equality.

Theorem 1.6 (Lück) *Let X be a connected compact CW complex whose fundamental group $G = \pi_1 X$ is residually finite. Then for every residual chain (G_i) in G and every $n \geq 0$ we have*

$$\lim_{i \to \infty} \frac{b_n(\overline{X}_i)}{[G : G_i]} = b_n^{(2)}(\widetilde{X}).$$

The proof of Lück's approximation theorem uses spectral calculus, a chapter within functional analysis of intrinsic beauty. We will thoroughly explain the ideas of this field in a preparatory section right before we give the proof of Lück's theorem. As explained above, the theorem can be restated as

$$\lim_{i \to \infty} b_n^{(2)}(\overline{X}_i) = b_n^{(2)}(\widetilde{X}).$$

It makes sense to ask if this equality remains true after dropping the assumption that G_i would have finite index in G. This leads to the *approximation conjecture* which, in a slightly weakened version, reads as follows.

Conjecture 1.7 (Approximation Conjecture) Let X be a connected compact CW complex, set $G = \pi_1 X$ and let (G_i) be a nested chain of normal subgroups of G with $\bigcap_i G_i = \{1\}$. Then for every $n \geq 0$ we have

$$\lim_{i \to \infty} b_n^{(2)}(\overline{X}_i) = b_n^{(2)}(\widetilde{X}).$$

The approximation conjecture has a decisive advantage over the approximation theorem: it allows for progress on the Atiyah conjecture and hence gives more positive results on Kaplansky's Conjecture 1.1. How? If one finds a chain (G_i) of normal subgroups in G with $\bigcap_i G_i = \{1\}$ such that the quotient groups G/G_i are torsion-free and satisfy the Atiyah conjecture, then the sequence $b_n^{(2)}(\overline{X}_i)$ consists of integers. So the limit $b_n^{(2)}(\widetilde{X})$ is an integer, too, hence G satisfies the Atiyah conjecture. In this sense, the class of torsion-free groups satisfying the Atiyah conjecture is *residually closed*.

Trying to transfer the proof of Lück's theorem to the approximation conjecture leads naturally to Schick's *determinant conjecture*. We will prove the determinant conjecture for residually finite groups which in turn shows the approximation conjecture for chains with residually finite factor groups G/G_i. This improves Lück's theorem from finite quotients to residually finite quotients. Many more variants and generalizations of Lück's theorem were meanwhile proven on which we will report in the course.

Lück's approximation theorem can be seen as a fundamental result in the active research field of *homology growth*: a positive n-th ℓ^2-Betti number of $\widetilde{X}$ detects that, asymptotically, the rank of the free part in the n-th homology of $\overline{X}_i$ grows linearly in the number of sheets. As a consequence of Lück's approximation, the following conjecture is a formally weaker version of the Singer conjecture.

Conjecture 1.8 Let X be an aspherical, $2n$-dimensional, closed, connected manifold with residually finite fundamental group $G = \pi_1 X$. Then for every residual chain (G_i) in G we have

$$\lim_{i \to \infty} \frac{\mathrm{rank}_{\mathbb{Z}} H_n(\overline{X}_i)_{\mathrm{free}}}{[G : G_i]} = (-1)^n \chi(X).$$

So the Euler characteristic is expected to detect *free* homology growth in the middle degree of an *even*-dimensional aspherical manifold. It so turns out that the aforementioned ℓ^2-torsion $\rho^{(2)}(\widetilde{X})$ serves as an odd-dimensional cousin of $\chi(X)$: it is expected to detect *torsion* homology growth in the middle degree of an *odd*-dimensional aspherical manifold.

Conjecture 1.9 Let X be an aspherical, $(2n + 1)$-dimensional, closed, connected manifold with residually finite fundamental group $G = \pi_1 X$. Then for every residual chain (G_i) in G we have

$$\lim_{i \to \infty} \frac{\log |H_n(\overline{X}_i)_{\mathrm{tors}}|}{[G : G_i]} = (-1)^n \rho^{(2)}(\widetilde{X}).$$

Note that the logarithm appearing in the formula says that non-zero ℓ^2-torsion actually detects *exponential* growth of the order of the torsion subgroup of $H_n(\overline{X}_i) = H_n(\overline{X}_i; \mathbb{Z})$. The definition of ℓ^2-torsion is somewhat involved. Conceptually, though, it is simply the ℓ^2-counterpart to classical Reidemeister torsion which once gave the complete classification of *Lens spaces*. We will explain this background before giving the precise definition of ℓ^2-torsion, followed by basic properties and some applications. Then we discuss that a potential proof of Conjecture 1.9 would actually split into three proofs of three different conjectures, each of which is of independent interest, and each of which is wide open: the *torsion Singer conjecture*, the *small regulator conjecture*, and the *determinant approximation conjecture*.

A typical situation of both geometric and algebraic interest arises if the odd-dimensional aspherical manifold X is a so-called *arithmetic locally symmetric space*. For example, G could be a finite index torsion-free subgroup of $SL(3; \mathbb{Z})$ and X would be the double coset space $G\backslash SL(3; \mathbb{R})/SO(3)$. This explicit example does not quite meet the requirements of Conjecture 1.9 because X is not compact. It is however homotopy equivalent to a compact CW complex so that one may still hope the conclusion of Conjecture 1.9 was true for residual chains (G_i) in G. Carrying out all computations, we would then obtain the remarkable formula

$$\lim_{i \to \infty} \frac{\log |H_2(\overline{X}_i)_{\text{tors}}|}{[G : G_i]} = \frac{\zeta(3)}{96\sqrt{3}\pi^2}$$

with $\zeta(3) = \sum_{n \geq 1} \frac{1}{n^3}$. While this must remain conjectural, some definite results are possible for compact arithmetic locally symmetric spaces if one replaces the $\mathbb{Z}$-coefficients in $H_n(\overline{X}_i) = H_n(\overline{X}_i; \mathbb{Z})$ with certain coefficient systems coming from representations of the matrix group in which (G) lies. This approach is due to Bergeron and Venkatesh and shall be presented in one of the more advanced sections of this text.

ℓ^2-Torsion has recently also come into focus in 3-manifold theory, where an additional twist in the definition leads to the ℓ^2-*Alexander torsion*. The ℓ^2-Alexander torsion of a 3-manifold determines the *Thurston norm* which in turn has played a central role in the recent breakthrough proof of the *virtually fibered conjecture*. Moreover, just like in the case of homology, both ℓ^2-Betti numbers and ℓ^2-torsion cannot only be defined for CW complexes but also for groups via *classifying spaces*. As such they define powerful tools to study groups and many interesting new questions arise. To name one, one could ponder how much information on the ℓ^2-torsion of a group is already contained in the *profinite completion* of the group? We will see, for example, that if we knew that Conjecture 1.9 was true, we could conclude that the ℓ^2-torsion and hence the volume of 3-manifolds depends on the profinite completion of the fundamental group only. In another vein, it is a consequence of Lück's approximation theorem that the first ℓ^2-Betti number $b_1^{(2)}(G)$ of a finitely presented residually finite group G captures the growth of the free abelian rank of the abelianized finite index subgroups of G. This turns $b_1^{(2)}(G)$ into a particularly relevant group invariant. It is related to M. Lackenby's *rank gradient* and D. Gaboriau's *cost* by the inequalities

$$\text{RG}(G) \geq \text{cost}(G) - 1 \geq b_1^{(2)}(G).$$

The second inequality is due to Gaboriau while the first is elegantly explained by the *Abért–Nikolov theorem*. It is a standing question if either inequality can be strict.

With varying resolution by details, all these aspects will be discussed during the course. This means the text at hand intends to set up a walkable path from the definition of Hilbert space to the state of the art in some specific questions. Operator algebras, Hilbert modules and von Neumann dimension will be introduced and discussed in Chap. 2 which assumes no prior knowledge on functional analysis. The study of ℓ^2-Betti numbers of CW complexes is the subject of Chap. 3. Chapter 4 introduces Lück's extended von Neumann dimension and ℓ^2-Betti numbers of groups via classifying spaces. Chapter 5 is concerned with the approximation of ℓ^2-Betti numbers, rank gradient, and cost. In Chap. 6 we study ℓ^2-*torsion*, torsion growth in twisted and untwisted homology and applications. Most sections in the first few chapters end with a number of problems which are meant to familiarize the reader with the acquired material and give an opportunity to try out the new methods in practice. The exercises vary in difficulty between almost obvious and pretty involved; but they have all been tested in practice. Needless to say: doing them is crucial.

Chapter 2
Hilbert Modules and von Neumann Dimension

2.1 Hilbert Spaces

Euclidean geometry is the geometry of $\mathbb{R}^n$. To talk about *lengths*, *angles* and *orthogonality* in a general finite dimensional $\mathbb{R}$-vector space V, we thus have to fix an identification $\psi \colon V \xrightarrow{\sim} \mathbb{R}^n$ first. On a second thought, this demands more than necessary because if two identifications ψ_1, ψ_2 differ by an orthogonal transformation $\psi_1 \circ \psi_2^{-1} \in O(n)$, then all lengths and angles agree and we are dealing with one and the same geometry on V. So what we really have to pick is a basis up to orthogonal transformations or, which is the same, an *inner product* on V: a positive definite, bilinear, symmetric form. The correct way of extending this notion from real to complex, possibly infinite-dimensional vector spaces is captured by the following definition.

Definition 2.1 An *inner product space* is a complex vector space V with a function $\langle \cdot, \cdot \rangle \colon V \times V \to \mathbb{C}$ which satisfies for all $x, y, z \in V$ and $\alpha \in \mathbb{C}$

- (i) $\langle x, x \rangle \geq 0$ with equality if and only if $x = 0$,
- (ii) $\langle x, y + z \rangle = \langle x, y \rangle + \langle x, z \rangle$,
- (iii) $\langle x, \alpha y \rangle = \alpha \langle x, y \rangle$,
- (iv) $\langle y, x \rangle = \overline{\langle x, y \rangle}$.

With this definition the inner product is conjugate-linear in the first variable and linear in the second variable. This stipulation appears to be more common in physics than in mathematics but we prefer to have the complication up front.

Example 2.2 Standard complex n-space $\mathbb{C}^n$ with standard inner product

$$\langle x, y \rangle = \sum_{i=1}^{n} \overline{x_i} y_i$$

© Springer Nature Switzerland AG 2019
H. Kammeyer, *Introduction to ℓ^2-invariants*, Lecture Notes in Mathematics 2247,
https://doi.org/10.1007/978-3-030-28297-4_2

for $x = (x_1, \ldots, x_n)$, $y = (y_1, \ldots, y_n) \in \mathbb{C}^n$.

Example 2.3 Complex valued continuous functions on an interval $C[a, b]$ with inner product

$$\langle f, g \rangle = \int_a^b \overline{f(x)} g(x) \mathrm{d}x$$

for $f, g \in C[a, b]$.

Let $(V, \langle \cdot, \cdot \rangle)$ be an inner product space. Two vectors $x, y \in V$ are called *orthogonal* if $\langle x, y \rangle = 0$. A subset $\{x_i \mid i \in I\} \subset V$ is called *orthonormal* if $\langle x_i, y_i \rangle = 1$ for all $i \in I$ and $\langle x_i, y_j \rangle = 0$ for $i \neq j$. We set $\|x\| = \sqrt{\langle x, x \rangle}$ and thus commit ourselves to verifying in what follows that this defines a norm on V. As a first step we observe a "Pythagorean theorem".

Lemma 2.4 *Let $\{x_i\}_{i=1}^n$ be orthonormal in V. Then for all $x \in V$ we have*

$$\|x\|^2 = \sum_{i=1}^n |\langle x_i, x \rangle|^2 + \left\| x - \sum_{i=1}^n \langle x_i, x \rangle x_i \right\|^2.$$

Proof This is an easy calculation after decomposing x orthogonally into

$$x = \sum_{i=1}^n \langle x_i, x \rangle x_i + \left(x - \sum_{i=1}^n \langle x_i, x \rangle x_i \right). \qquad \square$$

In particular, we obtain $\|x\|^2 \geq \sum_{i=1}^n |\langle x_i, x \rangle|^2$ which is sometimes known as *Bessel's inequality*.

Corollary 2.5 (Cauchy–Schwarz Inequality) *For all $x, y \in V$ we have $|\langle x, y \rangle| \leq \|x\| \|y\|$.*

Proof If $y = 0$ there is nothing to prove. Otherwise $\{\frac{y}{\|y\|}\}$ is an orthonormal set so that Bessel's inequality gives $\|x\|^2 \geq |\langle \frac{y}{\|y\|}, x \rangle|^2 = \frac{|\langle x, y \rangle|^2}{\|y\|^2}$. $\qquad \square$

Lemma 2.6 *The pair $(V, \|\cdot\|)$ is a normed vector space.*

Proof Only the triangle inequality needs proof. For $x, y \in V$ we calculate

$$\|x + y\|^2 = \langle x, x \rangle + \langle x, y \rangle + \langle y, x \rangle + \langle y, y \rangle = \langle x, x \rangle + 2 \operatorname{Re}\langle x, y \rangle + \langle y, y \rangle$$

$$\leq \langle x, x \rangle + 2|\langle x, y \rangle| + \langle y, y \rangle \leq \|x\|^2 + 2\|x\|\|y\| + \|y\|^2$$

$$= (\|x\| + \|y\|)^2$$

by the Cauchy–Schwarz inequality. $\qquad \square$

As usual, a normed space is a metric space with respect to the distance $d(x, y) = \|x - y\|$. Recall that a metric space X is called *complete* if every Cauchy-sequence of points in X has a limit in X.

Definition 2.7 A *Hilbert space* is an inner product space which is complete as metric space.

Standard n-space $\mathbb{C}^n$ from Example 2.2 is a Hilbert space. The direct sum $\bigoplus_{k=1}^{\infty} \mathbb{C}$ clearly inherits an inner product from the inclusions of the subspaces $\mathbb{C}^n = \bigoplus_{k=1}^{n} \mathbb{C}$ but it is not complete because $\left(\sum_{k=1}^{N} \frac{e_k}{k^2}\right)_N$ is a Cauchy sequence without limit (here e_k is the k-th standard basis vector). However, any inner product space can be transformed into a Hilbert space by *completion*.

Example 2.8 Let ℓ^2 be the space of sequences $(a_n)_{n=0}^{\infty}$ of complex numbers which satisfy $\sum_{n=0}^{\infty} |a_n|^2 < \infty$ with the inner product

$$\langle (a_n), (b_n) \rangle = \sum_{n=0}^{\infty} \overline{a_n} b_n.$$

The series converges because $2|a_n||b_n| \leq |a_n|^2 + |b_n|^2$. One can check that l^2 is a Hilbert space in which $\bigoplus_{k=1}^{\infty} \mathbb{C}$ embeds isometrically and densely. Similarly, $C[a, b]$ from Example 2.3 is not a Hilbert space (why?). Its completion looks as follows.

Example 2.9 Let $\mathcal{L}^2[a, b]$ be the complex vector space of complex valued Lebesgue measurable functions on the interval $[a, b]$ which satisfy $\int_a^b |f|^2 d\lambda < \infty$. Setting

$$\langle f, g \rangle = \int_a^b \overline{f} g \, d\lambda$$

defines an inner product on $\mathcal{L}^2[a, b]$ where convergence follows again because $2|f||g| \leq |f|^2 + |g|^2$. Let $L^2[a, b]$ be the quotient space of $\mathcal{L}^2[a, b]$ by the subspace of functions which vanish Lebesgue-almost everywhere (or, which is the same, the functions f for which $\|f\| = 0$). Then the inner product of $\mathcal{L}^2[a, b]$ descends to an inner product on $L^2[a, b]$ and turns $L^2[a, b]$ into a Hilbert space in which $C[a, b]$ embeds isometrically and densely.

We can form new Hilbert spaces out of old ones as follows.

1. If H is a Hilbert space, then so is every closed subspace $K \subset H$. The example $C[a, b] \subset L^2[a, b]$ shows that the word "closed" cannot be omitted.
2. Similarly, if $K \subset H$ is a closed subspace, then the quotient space H/K is a Hilbert space because the canonical map $K^{\perp} \to H/K$ identifies H/K with the *orthogonal complement*

$$K^{\perp} = \{x \in H \mid \langle x, y \rangle = 0 \text{ for all } y \in K\}$$

of K which is a closed subspace. The identification is more subtle than it sounds because for constructing the inverse map $H/K \to K^\perp$ one has to show that every affine space $(x + K) \in H/K$ has a unique element of minimal norm (Exercise 2.1.2).

3. If $\{H_i\}_{i=1}^\infty$ is an (at most) countable family of Hilbert spaces, then we define the *direct sum* $\bigoplus_{i=1}^\infty H_i$ as the space of all sequences $(x_i)_{i=0}^\infty$ with $x_i \in H_i$ satisfying $\sum_{i=1}^\infty \|x_i\|_{H_i}^2 < \infty$. The inner products of the H_i sum up (independent of order) to give an inner product on $\bigoplus_{i=1}^\infty H_i$ for which $\bigoplus_{i=1}^\infty H_i$ is complete. Note that if the family is finite, the condition $\sum_{i=1}^\infty \|x_i\|_{H_i}^2 < \infty$ is empty.

4. For two Hilbert spaces H_1 and H_2 we declare an inner product on the vector space tensor product $H_1 \otimes_{\mathbb{C}} H_2$ by setting

$$\langle x_1 \otimes x_2, y_1 \otimes y_2 \rangle_{H_1 \otimes_{\mathbb{C}} H_2} = \langle x_1, y_1 \rangle_{H_1} \langle x_2, y_2 \rangle_{H_2}$$

on simple tensors. This extends linearly to all of $H_1 \otimes_{\mathbb{C}} H_2$ by writing a general element in $H_1 \otimes_{\mathbb{C}} H_2$ as a sum (no coefficients!) of simple tensors. We define the Hilbert space *tensor product* $H_1 \otimes H_2$ as the Hilbert space completion of the inner product space $(H_1 \otimes_{\mathbb{C}} H_2, \langle \cdot, \cdot \rangle_{H_1 \otimes_{\mathbb{C}} H_2})$. As an example, given measure spaces (X_i, μ_i) with countably generated σ-algebras for $i = 1, 2$, the correspondence $f \otimes g \mapsto f \cdot g$ extends to a canonical identification

$$L^2(X_1, \mathrm{d}\mu_1) \otimes L^2(X_2, \mathrm{d}\mu_2) \cong L^2(X_1 \times X_2, \mathrm{d}\mu_1 \otimes \mathrm{d}\mu_2).$$

For a proof we refer to [150, Theorem II.10(a), p. 52].

In linear algebra we were taught that as an application of Zorn's lemma, every vector space has a basis. In the context of Hilbert spaces such a basis is sometimes referred to as a *Hamel basis* in order to distinguish it from the following concept which is more natural and more convenient in our setting.

Definition 2.10 Let H be a Hilbert space. An *orthonormal basis* of H is a maximal orthonormal subset of H.

Here, as usual, "maximal" means that the orthonormal set is not properly contained in any other orthonormal set. Existence is again immediate from Zorn's lemma because the union of a nested sequence of orthonormal subsets is orthonormal. The next result says that the orthogonal decomposition of vectors familiar from finite-dimensional Euclidean space carries over to Hilbert spaces though the sums are possibly infinite.

Theorem 2.11 *Let H be a Hilbert space, let $\{x_i\}_{i \in I}$ be an orthonormal basis and let $x \in H$. Then*

$$x = \sum_{i \in I} \langle x_i, x \rangle x_i \quad and \quad \|x\|^2 = \sum_{i \in I} |\langle x_i, x \rangle|^2.$$

It follows from Bessel's inequality $\sum_{j \in J} |\langle x_j, x \rangle|^2 \leq \|x\|^2$ for $J \subset I$ finite that $\langle x_i, x \rangle$ is nonzero for at most countably many $i \in I$. The theorem asserts convergence of the sums in H and $\mathbb{R}$ independent of order.

Proof Fix some ordering $x_{i_1}, x_{i_2}, \ldots$ of the basis vectors x_i with $\langle x_i, x \rangle$ nonzero and set $y_n = \sum_{k=1}^{n} \langle x_{i_k}, x \rangle x_{i_k}$. For $n > m$ we get

$$\|y_n - y_m\|^2 = \left\| \sum_{k=m+1}^{n} \langle x_{i_k}, x \rangle x_{i_k} \right\|^2 = \sum_{k=m+1}^{n} |\langle x_{i_k}, x \rangle|^2 .$$

This shows that y_n is a Cauchy sequence because the series $\sum_{k=1}^{\infty} |\langle x_{i_k}, x \rangle|^2$ converges since the partial sums form a monotone increasing sequence bounded by $\|x\|^2$. Using orthonormality of $\{x_i\}_{i \in I}$ it is easy to see that $y = \lim_{n \to \infty} y_n$ equals x. By Lemma 2.4 we have moreover

$$0 = \lim_{n \to \infty} \|x - \sum_{k=1}^{n} \langle x_{i_k}, x \rangle x_{i_k}\|^2 = \|x\|^2 - \sum_{k=1}^{\infty} |\langle x_{i_k}, x \rangle|^2$$

which shows the second equality. □

Most Hilbert spaces of interest possess a countable orthonormal basis. In that case we say that the Hilbert space H is *separable*. An equivalent characterization of separability is the existence of a countable dense subset in H (Exercise 2.1.6). We remark that for a separable Hilbert space an orthonormal basis can be constructed by a Gram–Schmidt procedure (Exercise 2.1.5) without invoking the axiom of choice. In this course we will exclusively deal with separable Hilbert spaces.

A morphism $A: H_1 \to H_2$ of Hilbert spaces is a *bounded linear operator*. This means that $A(\mu_1 x + \mu_2 y) = \mu_1 A x + \mu_2 A y$ for all $x, y \in H_1$ and $\mu_1, \mu_2 \in \mathbb{C}$ and that there exists a constant $C \geq 0$ with $\|Ax\|_{H_2} \leq C\|x\|_{H_1}$ for all $x \in H_1$. A bounded linear operator is by definition (Lipschitz) continuous. If A is a continuous linear operator, then there exists $\delta > 0$ such that $\|Ax\| \leq 1$ for $\|x\| < \delta$. Thus for any nonzero $y \in H_1$ we have $\left\| A\left(\frac{\delta y}{2\|y\|} \right) \right\| < 1$, so $C = \frac{2}{\delta}$ is a constant showing that A is bounded. So a bounded linear operator is the same as a continuous linear operator. We stress that morphisms of Hilbert spaces are not required to preserve the inner product. If they do, that is if $\langle Ux, Uy \rangle_{H_2} = \langle x, y \rangle_{H_1}$, then U is called an *isometry*. If in addition U is onto, then $U: H_1 \to H_2$ is called *unitary* and H_1 and H_2 are called *isomorphic*.

Theorem 2.12 *Any separable Hilbert space H is either isomorphic to $\mathbb{C}^n$ or to ℓ^2.*

Proof Choose an orthonormal basis $\{x_i\}_{i \in I}$ of H. Then I can either be identified with $\{1, \ldots, n\}$ or with the set of all positive integers. Accordingly, sending $x \in H$ to $(\langle x_1, x \rangle, \langle x_2, x \rangle, \ldots)$ defines either a map from H to $\mathbb{C}^n$ or, by Theorem 2.11, to ℓ^2. This map is clearly linear and continuous. Theorem 2.11 shows moreover that it is norm preserving and the methods of the proof also give that it is onto. By polarization (Exercise 2.1.1) it is unitary. □

Remark 2.13 It should be a reassuring fact that all infinite-dimensional separable Hilbert spaces are isomorphic. The experience, however, is that this observation causes some headache to anyone learning this for the first time. For in a moment we will discuss that also $L^2[a, b]$ is separable, so why then, oh why, do we give all these fancy names $L^2[a, b]$, ℓ^2, $\ell^2 G$ to one and the same Hilbert space? The reason is that we are not only interested in the abstract Hilbert space on its own but more so in *representations* of various algebraic and functional analytic objects on Hilbert space. To even write down any such natural representations we need to give separable Hilbert space its suitable interpretation as square-integrable functions, square-summable sequences, and so forth.

Example 2.14 The most important so obtained identification of Hilbert spaces, both historically and for the theory of ℓ^2-invariants, is the isomorphism $L^2[-\pi, \pi] \cong \ell^2(\mathbb{Z})$ called *Fourier transform*. Theorem 2.12 implements this isomorphism as soon as we have found a countable orthonormal basis of $L^2[-\pi, \pi]$. We claim that $\left\{ f_n(x) = \frac{e^{inx}}{\sqrt{2\pi}} \right\}_{n=-\infty}^{\infty}$ is such an orthonormal basis. It is clearly an orthonormal set so we only need to show maximality which is equivalent to showing that $\langle f_n, g \rangle = 0$ for all $n \in \mathbb{Z}$ implies $g = 0$. To show the latter, the following result is key.

Theorem 2.15 *Let f be a 2π-periodic continuously differentiable function. Then the sequence of functions $\sum_{n=-N}^{N} \langle f_n, f \rangle f_n$ converges uniformly to f.*

The proof is involved and goes beyond the scope of this chapter; one has to show Cesàro summability of the series first to conclude uniform convergence by some estimates also involving the derivative f', see [150, Problems II.14, II.15, p. 64]. Now for any f as in the theorem, $\sum_{n=-N}^{N} \langle f_n, f \rangle f_n$ converges uniformly and thus also in $L^2[-\pi, \pi]$. So if $\langle f_n, g \rangle = 0$ for all $n \in \mathbb{Z}$, then also

$$\langle f, g \rangle = \left\langle \sum_{n=-\infty}^{\infty} \langle f_n, f \rangle f_n, g \right\rangle = \sum_{n=-\infty}^{\infty} \langle f_n, f \rangle \langle f_n, g \rangle = 0.$$

Thus g lies in the orthogonal complement of the continuously differentiable periodic functions $C_p^1[-\pi, \pi]$. But $C_p^1[-\pi, \pi]$ is dense in $L^2[-\pi, \pi]$ because every step function is an L^2-limit of functions in $C_p^1[-\pi, \pi]$ and step functions are dense. This completes the proof. Since the interval $[-\pi, \pi]$ with normalized Lebesgue measure is isomorphic, as measure space, to the circle S^1 with standard rotation invariant Borel probability measure, we can equally interpret Fourier transform as the isomorphism of Hilbert spaces $L^2(S^1) \cong \ell^2(\mathbb{Z})$.

If for $f \in L^2[-\pi, \pi]$ we set $c_n = \langle f_n, f \rangle$, then the equalities in Theorem 2.11 take the form

$$f = \frac{1}{\sqrt{2\pi}} \sum_{n=-\infty}^{\infty} c_n e^{inx} \quad \text{and} \quad \sum_{n=-\infty}^{\infty} |c_n|^2 = \frac{1}{2\pi} \int_{-\pi}^{\pi} |f(x)|^2 dx.$$

The first one is called the *Fourier series* presentation of f with *Fourier coefficients* c_n and the second one is known as *Parseval's identity*. Motivated by Theorem 2.12, this terminology is also common usage in the abstract setting of Theorem 2.11. Note that Parseval's identity can be seen as the case "$n = \infty$" of our Pythagorean Lemma 2.4.

Exercises

2.1.1 Let $(V, \langle \cdot, \cdot \rangle)$ be an inner product space and let $x, y \in V$.

(a) Show the *parallelogram identity* $\|x + y\|^2 + \|x - y\|^2 = 2\|x\|^2 + 2\|y\|^2$.
(b) Show that the inner product can be recovered from the norm by *polarization* according to the formula

$$\langle x, y \rangle = \tfrac{1}{4}((\|x + y\|^2 - \|x - y\|^2) - i(\|x + iy\|^2 - \|x - iy\|^2)).$$

2.1.2 Let H be a Hilbert space, $K \subseteq H$ a closed subspace and $x \in H$. Show that there is a unique element $z \in K$ closest to x.

Hint: Choose a sequence (y_n) in K realizing $\inf_{y \in K} \|x - y\|$ and show that it is Cauchy. Exercise 2.1.1 might help. Don't forget uniqueness.

2.1.3 Two measures μ_1 and μ_2 on the same measurable space X are called *mutually singular* if there exists a measurable set $A \subset X$ such that $\mu_1(A) = 0$ and $\mu_2(X \setminus A) = 0$. Let μ_1 and μ_2 be two mutually singular Borel measures on the real line. Show that $L^2(\mathbb{R}, d(\mu_1 + \mu_2))$ is canonically isomorphic to $L^2(\mathbb{R}, d\mu_1) \oplus L^2(\mathbb{R}, d\mu_2)$.

2.1.4 Let H be a Hilbert space and $K \subseteq H$ a closed subspace. Show that every element $x \in H$ decomposes uniquely as $x = z + w$ where $z \in K$ and $w \in K^\perp = \{y \in H : \langle y, z \rangle = 0 \text{ for all } z \in K\}$.

2.1.5 Let H be a Hilbert space. Recall the Gram-Schmidt process for constructing an orthonormal set $v_1, v_2, \ldots \in H$ from an arbitrary sequence of vectors $u_1, u_2, \ldots \in H$: set $v_n = u_n - \sum_{k<n} \langle v_k, u_n \rangle$ and normalize. Apply it to the sequence $1, x, x^2, x^3, \ldots$ of functions in $L^2[-1, 1]$ and show that you obtain the sequence $p_n(x) = \sqrt{n + \tfrac{1}{2}} P_n(x)$ for $n = 0, 1, 2, \ldots$ where

$$P_n(x) = \frac{1}{2^n n!} \frac{d}{dx}^n (x^2 - 1)^n$$

is the n-th *Legendre polynomial*.

2.1.6 Show that a Hilbert space is separable if and only if it possesses a countable dense subset.

2.2 Operators and Operator Algebras

Let H and K be Hilbert spaces. We denote the set of morphisms from H to K by $B(H, K)$. Recall that a continuous bijection of topological spaces need not be a homeomorphism. Fortunately, there is no corresponding phenomenon for morphisms $T \in B(H, K)$.

Theorem 2.16 (Inverse Mapping) *If T is bijective, then T is invertible.*

If T is only surjective, then the theorem shows that T is open by factorizing T over $H/\ker T$. If we knew conversely that a surjective operator was open, we would get that the inverse map of a bijective $T \in B(H, K)$ is continuous. So the above can be restated as follows.

Theorem 2.17 (Open Mapping) *If T is onto, then T is open.*

We shall take the open mapping theorem for granted; a suitable reference is [150, Theorem III.10, p. 82]. The letter "B" in $B(H, K)$ is meant to remind us that any $T \in B(H, K)$ is required to be bounded which says there is $C \geq 0$ such that $\|Tx\| \leq C\|x\|$ for all $x \in H$. The minimal such C is called the *operator norm* of T or for short just the *norm* of T. It is customary to also denote it by $\|T\|$ and we have

$$\|T\| = \sup_{\|x\|_H = 1} \|Tx\|_K.$$

It thus follows from the norm properties of $\| \cdot \|_K$ that the operator norm is indeed a norm on the complex vector space $B(H, K)$ and the induced topology is called the *uniform operator topology* or simply the *norm topology*. One can show that the normed space $B(H, K)$ is complete and thus by definition a *Banach space*. Two cases are of particular interest: the *dual space* $H^* = B(H, \mathbb{C})$ and the endomorphisms $B(H) = B(H, H)$ better known as the *bounded operators* on H.

For many purposes, the norm topology on $B(H)$ has too many open sets (is *too fine*), so that subsets of interest in $B(H)$ have too small closures. That is why one introduces two coarser topologies. The *strong operator topology* is the coarsest topology in which all evaluation maps $E_x : B(H) \to H$, $T \mapsto Tx$ for $x \in H$ are still continuous. The *weak operator topology* is the coarsest topology for which all the maps $E_{x,y} : B(H) \to \mathbb{C}$, $T \mapsto \langle x, Ty \rangle$ for $x, y \in H$ are continuous. If H is infinite dimensional, neither the weak nor the strong operator topologies are first countable. This has the effect that sequences need to be replaced by *nets* to describe closures of subsets. A *net* in a topological space X is a map $I \to X$ from a *directed set* $(I, \leq)$ where "directed" means it comes with a reflexive and transitive binary relation "$\leq$" such that any two elements $a, b \in I$ have a *common upper bound* $c \in I$ with $a \leq c$ and $b \leq c$. A net $(x_i)_{i \in I}$ in X *converges* to $x \in X$ if for each neighborhood U of x there is $i \in I$ such that $x_j \in U$ for all $j \geq i$. It is then true for a completely general topological space X that a subset $A \subset X$ is closed if and only if all nets in A which are convergent in X have all their limits in A. It is likewise true that a map $f : X \to Y$ of arbitrary topological spaces is continuous at $x \in X$

if and only if $\lim_{i \in I} f(x_i) = f(x)$ for all nets $(x_i)_{i \in I}$ in X converging to x. The following result explains why one does not encounter the dual Hilbert space H^* too often in writings.

Theorem 2.18 (Riesz Lemma) *Given $T \in H^*$ there exists a unique $y_T \in H$ such that $T(x) = \langle y_T, x \rangle$. Moreover, we have $\|T\| = \|y_T\|_H$.*

Proof We only give an instructional outline. If $T = 0$, then $y_T = 0$ is the unique vector doing the trick. Otherwise, there exists $z \in H \backslash \ker(T)$. Since T is continuous, $\ker(T)$ is closed so that by Exercise 2.1.2, there is a unique $u \in \ker(T)$ closest to z. One checks that $w = z - u \in (\ker T)^{\perp}$ and that $y_T = \frac{\overline{Tw}}{\|w\|^2} w$ is the unique element as desired. Details can be found in [150, Theorem II.4, p. 43]. □

Given $T \in B(H, K)$, we obtain the *adjoint* $T^* \in B(K, H)$ by setting $T^*x = y_{l(T,x)}$ where $l(T, x) \in H^*$ is the linear functional $h \mapsto \langle x, Th \rangle_K$. We thus have enforced the characterizing equality $\langle T^*x, y \rangle = \langle x, Ty \rangle$. From this it follows that

$$\ker T^* = (\operatorname{im} T)^{\perp} \quad \text{or equivalently} \quad (\ker T^*)^{\perp} = \overline{\operatorname{im} T}$$

where the bar means closure. If T has a bounded inverse T^{-1}, then so does T^* and $(T^*)^{-1} = (T^{-1})^*$. By means of adjoints unitaries and isometries can be conveniently characterized.

- An operator $U \in B(H, K)$ is unitary if and only if $U^*U = \operatorname{id}_H$ and $UU^* = \operatorname{id}_K$.
- An operator $U \in B(H, K)$ is an isometry if and only if $U^*U = \operatorname{id}_H$.
- A still weaker notion is that of a *partial isometry* $U \in B(H, K)$ where we only require that U^*U is a *projection* in $B(H)$.
- An (orthogonal) *projection* is an operator $P \in B(H)$ satisfying $P = P^* = P^2$. Geometrically, P is the orthogonal projection onto $\operatorname{im} P$ because $P^2 = P$ implies $H = \operatorname{im} P \oplus \ker P$ and $P = P^*$ implies $\operatorname{im} P \subset (\ker P)^{\perp}$, hence $\operatorname{im} P = (\ker P)^{\perp}$ by Exercise 2.1.4. If an operator $P \in B(H)$ satisfies only $P^2 = P$ but not $P = P^*$, we will explicitly call it an *oblique projection*.
- A projection is an example of a *positive* operator, an operator $A \in B(H)$ satisfying $\langle Ax, x \rangle \geq 0$ for all $x \in H$. We write $A \leq B$ for $A, B \in B(H)$ if $B - A$ is positive.
- Every positive operator is in particular a *self-adjoint operator*: an operator $T \in B(H)$ such that $T = T^*$. This is an easy exercise, namely Exercise 2.2.4 (ii).
- Finally, a self-adjoint operator is a special kind of a *normal operator*: an operator $T \in B(H)$ which commutes with its adjoint, $T^*T = TT^*$.

For $H = K$ the correspondence $T \mapsto T^*$ defines an *involution*: a conjugate linear, norm preserving bijection of $B(H)$ satisfying $(TS)^* = S^*T^*$ and $(T^*)^* = T$. A short form of saying this is that $B(H)$ is a **-algebra* (read "star algebra"). The *commutant* of a subset $M \subset B(H)$ is given by $M' = \{T \in B(H) \mid ST = TS \text{ for all } S \in M\}$, the *bicommutant* is $M'' = (M')'$.

Theorem 2.19 (von Neumann Bicommutant Theorem) *Let M be a unital (meaning $\mathrm{id}_H \in M$) *-subalgebra of $B(H)$. The following are equivalent.*

  (i) M is weakly closed.
 (ii) M is strongly closed.
(iii) $M = M''$.

Proof Of course (i) $\Rightarrow$ (ii). To see (iii) $\Rightarrow$ (i) we show that commutants are always weakly closed. So let $N \subset B(H)$ be any subset. If $N' = B(H)$, we are done. Otherwise, there is $S \in N$, $T_0 \in B(H) \setminus N'$ and $x \in H$ such that $ST_0 x - T_0 Sx = y \neq 0$. So the map $T \mapsto \langle (ST - TS)x, y \rangle$ takes a nonzero value at $T = T_0$. But since this map is weakly continuous, it does so for an entire weakly open neighborhood U of T_0. Thus $U \subset B(H) \setminus N'$ which shows that N' is weakly closed. To see (ii) $\Rightarrow$ (iii) first note that the inclusion $M \subset M''$ is tautological. To obtain the other inclusion we observe that the strong operator topology is the topology of pointwise convergence in H so that a neighborhood basis of $T \in B(H)$ is given by

$$N(T; x_1, \ldots, x_n; \varepsilon) = \{S \in B(H) \colon \|(T - S)x_i\| < \varepsilon \text{ for all } i\}.$$

So given $T \in M''$, we want to find $S \in M$ within this neighborhood. Let $x = (x_1, \ldots, x_n) \in H^n$. The diagonal action of M on H^n embeds M in $B(H^n) = \mathrm{Mat}(n, n; B(H))$ as constant diagonal matrices. For this embedding, however, $M' = \mathrm{Mat}(n, n; M')$ which is why M'' consists again of constant diagonal matrices with constant entry in $M'' \subseteq B(H)$. So M'' is embedded in $B(H^n)$ the same way as M is. In what follows the vectors under consideration determine which embedding is meant. Let P be the orthogonal projection onto $K = \overline{Mx}$. We claim that $P \in M'$. Indeed, K is obviously M-invariant and so is $K^{\perp}$ because $M^* = M$. Decomposing any $y \in H^n$ uniquely as $y = y_K + y_{K^\perp}$ we get for every $A \in M$ that

$$PAy = P(Ay_K + Ay_{K^\perp}) = Ay_K = APy$$

hence the claim. Now M is unital, so $x \in Mx$ and thus $Tx = TPx = PTx \in K = \overline{Mx}$. Therefore there is $S \in M$ such that $\|Tx - Sx\| < \varepsilon$. In particular $\|Tx_k - Sx_k\| < \varepsilon$ for $k = 1, \ldots, n$. □

Definition 2.20 A unital *-subalgebra of $B(H)$ satisfying one (then all) of the above conditions is called a *von Neumann algebra*.

We remark that a norm closed *-subalgebra is known as a C^*-*algebra* (read "C-star algebra"). So every von Neumann algebra is a C^*-algebra and satisfies the so called C^* *identity* $\|T^*T\| = \|T\|^2$ which is easy to see for adjoints and can be used to characterize C^*-algebras abstractly.

Example 2.21 The trivial examples of von Neumann algebras are $\mathbb{C}$ (realized as multiples of id_H) and $B(H)$. Note that one is the commutant of the other.

The von Neumann algebra we are about to construct in the next example is key. It foreshadows that the fields of functional analysis and group theory share a vast overlap and exploring their mutual interaction remains an object of active research to this day. *Now and in the remainder of the text, unless otherwise stated, G will denote a discrete, countable group.*

Example 2.22 (Group von Neumann Algebra) The *group algebra* (or *group ring*) $\mathbb{C}G$ is the $\mathbb{C}$-vector space spanned by G with multiplication defined on the basis G by composition in the group and on $\mathbb{C}G$ by linear extension. Thus $\mathbb{C}G$ is a commutative algebra if and only if G is a commutative group. Requiring that G be orthonormal turns $\mathbb{C}G$ into an inner product space. In concrete terms, $\mathbb{C}G$ consists of finite formal sums $\sum_{g \in G} c_g g$ with distributive multiplication and inner product given by

$$\langle \textstyle\sum_g c_g g, \sum_g d_g g \rangle = \sum_g \overline{c_g} d_g.$$

The Hilbert space completion of $\mathbb{C}G$ is denoted by $\ell^2 G$. By construction $G \subset \ell^2 G$ is an orthonormal (Hilbert) basis. So elements of $\ell^2 G$ can be represented by Fourier series $\sum_{g \in G} c_g g$ with $\sum_{g \in G} |c_g|^2 < \infty$ as in Theorem 2.11. An element $h \in G$ acts unitarily on $\ell^2 G$ by $g \mapsto hg$ and also by $g \mapsto gh^{-1}$ for basis elements $g \in G$. By linear extension this defines the *left regular representation* λ and the *right regular representation* ρ of $\mathbb{C}G$, respectively, and turns $\ell^2 G$ into a $\mathbb{C}G$-bimodule. We embed the group algebra as bounded operators on $\ell^2 G$ by the right regular representation $\rho \colon \mathbb{C}G \hookrightarrow B(\ell^2 G)$. The *-operation restricts on $\rho(\mathbb{C}G)$ to the involution $\rho \left(\sum_{g \in G} c_g g \right) \mapsto \rho \left(\sum_{g \in G} \overline{c_g} g^{-1} \right)$.

Definition 2.23 The *group von Neumann algebra* $\mathcal{R}(G)$ is the weak closure of the unital *-subalgebra $\rho(\mathbb{C}G) \subset B(\ell^2 G)$.

By Theorem 2.19 the group von Neumann algebra $\mathcal{R}(G)$ is equivalently the strong closure of $\rho(\mathbb{C}G)$ or equivalently $\mathcal{R}(G) = \rho(\mathbb{C}G)''$. Whenever the distinction matters, we will say more precisely that $\mathcal{R}(G)$ is the *right group von Neumann algebra* of G. Exercise 2.2.9 provides a guided tour through the proof that the commutant $\mathcal{R}(G)'$ coincides with the *left group von Neumann algebra* $\mathcal{L}(G) = \lambda(\mathbb{C}G)''$ generated by the left regular embedding of $\mathbb{C}G$. As group multiplication is associative, left and right regular representation commute so that $\rho(\mathbb{C}G)$ lies in $B(\ell^2 G)^\lambda$ while $\lambda(\mathbb{C}G)$ lies in $B(\ell^2 G)^\rho$, the subalgebras of $B(\ell^2 G)$ consisting of left and right G-equivariant operators on $\ell^2 G$, respectively. It turns out that these inclusions are weakly dense.

Theorem 2.24 *We have $\mathcal{R}(G) = B(\ell^2 G)^\lambda$ and $\mathcal{L}(G) = B(\ell^2 G)^\rho$.*

Proof By symmetry it is enough to show the first equality. To see the inclusion "$\subseteq$" we only have to observe that $B(\ell^2 G)^\lambda$ is strongly closed. So if $(T_i)_{i \in I}$ is a net in $B(\ell^2 G)^\lambda$ converging strongly to $T \in B(\ell^2 G)$ and $x \in \ell^2 G$ is any vector, then for all $g \in G$ we have

$$T(gx) = E_{gx}(\lim_{i \in I} T_i) = \lim_{i \in I} E_{gx}(T_i) = \lim_{i \in I} gT_i(x) = g \lim_{i \in I} T_i(x) = gT(x)$$

thus $T \in B(\ell^2 G)^\lambda$. For the other inclusion we use that each $S \in \mathcal{R}(G)' = \lambda(\mathbb{C}G)''$ is a strong limit of a net in $\lambda(\mathbb{C}G)$. Hence every $T \in B(\ell^2 G)^\lambda$ commutes with every such S. This gives $B(\ell^2 G)^\lambda \subset \mathcal{R}(G)'' = \mathcal{R}(G)$. $\square$

Example 2.25 Suppose that G in the above example is a finite group of order n. Then $\ell^2 G \cong \mathbb{C}^n$ is a finite-dimensional Hilbert space, the various topologies on the $(n \times n)$-matrices $B(\ell^2 G) = M_n(\mathbb{C})$ agree and $\mathbb{C}G$ embeds as a closed subalgebra. Thus $\mathcal{R}(G) = \mathbb{C}G$ and in particular $\mathcal{R}(\{1\}) = \mathbb{C}$.

Example 2.26 Building upon Example 2.14 let us now consider the example $G = \mathbb{Z}$ of Example 2.22. The left action of the generator $1 \in \mathbb{Z}$ on $\ell^2(\mathbb{Z})$ shifts a basis element $k \in \mathbb{Z} \subset \ell^2(\mathbb{Z})$ by one step: $k \mapsto k + 1$. By Fourier transform this corresponds in $L^2[-\pi, \pi]$ to shifting a basis vector $\frac{e^{inx}}{\sqrt{2\pi}}$ to $\frac{e^{i(n+1)x}}{\sqrt{2\pi}} = e^{ix} \frac{e^{inx}}{\sqrt{2\pi}}$. Setting $z = e^{ix}$ and identifying $\mathbb{C}[\mathbb{Z}]$ with the Laurent polynomials $\mathbb{C}[z, z^{-1}]$, it follows that the left regular representation of $\mathbb{C}[\mathbb{Z}]$ on $\ell^2(\mathbb{Z})$ corresponds to multiplication of functions in $L^2[-\pi, \pi]$ by Laurent polynomials in $\mathbb{C}[z, z^{-1}]$. Thus $\mathcal{R}(\mathbb{Z}) = B(L^2[-\pi, \pi])^\lambda$ consists of operators which are equivariant with respect to Laurent polynomials. By the Stone–Weierstrass theorem every continuous function in $C[-\pi, \pi]$ is a uniform limit, thus an L^2-limit, of polynomials. Since $C[-\pi, \pi] \subset L^2[-\pi, \pi]$ is dense, it follows that any $f \in L^2[-\pi, \pi]$ is an L^2-limit of polynomials, $f = \lim_k p_k$. Since $T \in B(L^2[-\pi, \pi])^\lambda$ is continuous, we thus have

$$Tf = T(\lim_{k\to\infty} p_k) = \lim_{k\to\infty} Tp_k = \lim_{k\to\infty} T(p_k 1) = (\lim_{k\to\infty} p_k)T(1) = T(1) \cdot f,$$

so T is given by multiplication with the function $T(1) \in L^2[-\pi, \pi]$ where $1 \in L^2[-\pi, \pi]$ is the constant function. We claim that in fact $T(1)$ lies in the subspace $L^\infty[-\pi, \pi]$ of $L^2[-\pi, \pi]$ because T is *essentially bounded* by $\|T\|$. Suppose on the contrary the set $\{|T(1)| > \|T\|\}$, which is well-defined up to a null set, had positive Lebesgue measure. Then there is $\varepsilon > 0$ such that the set $A = \{|T(1)| \geq \|T\| + \varepsilon\}$ still has positive measure. Let χ_A be the *characteristic function* of A which is equal to one on A and zero elsewhere. Then the vector $f_A = \frac{\chi_A}{\sqrt{\lambda(A)}} \in L^2[-\pi, \pi]$ has norm one, so that we have

$$\|T\|^2 \geq \|T(f_A)\|^2 = \int_{-\pi}^{\pi} |T(1)|^2 \frac{\chi_A}{\lambda(A)} d\lambda \geq (\|T\| + \varepsilon)^2$$

which is absurd and proves the claim. Conversely, multiplication with any $g \in L^\infty[-\pi, \pi]$ clearly defines an element in $B(L^2[-\pi, \pi])^\lambda$. These constructions are mutually inverse. We thus have proven

$$\mathcal{R}(\mathbb{Z}) \cong L^\infty[-\pi, \pi] \cong L^\infty(S^1).$$

With absolutely no effort this result generalizes to $\mathcal{R}(\mathbb{Z}^n) \cong L^\infty(\mathbb{T}^n)$ where $\mathbb{T}^n$ is the n-dimensional torus.

Remark 2.27 The isomorphism $\mathcal{R}(\mathbb{Z}) \cong L^\infty(S^1)$ is only one appearance of a way more general principle: every *abelian* von Neumann algebra acting on a separable Hilbert space is isomorphic to $L^\infty(X, \mu)$ for some standard measure space (X, μ). Similarly, every unital *abelian* C^*-algebra is isomorphic to $C(X)$, the continuous functions on a compact Hausdorff space X. Isomorphism of abelian von Neumann algebras corresponds to isomorphism of measure spaces and isomorphism of abelian unital C^*-algebras corresponds to homeomorphism of compact Hausdorff spaces. So one might want to think about a noncommutative von Neumann algebra as a "noncommutative measure space" whereas a noncommutative C^*-algebra should be a "noncommutative topological space". The study of operator algebras is therefore frequently subsumed under the somewhat glamorous notion *noncommutative geometry*.

Example 2.28 For each positive integer n, we can amplify the group von Neumann algebra to $B\left((\ell^2 G)^n\right)^\lambda$ where G acts diagonally by λ. The discussion above Theorem 2.24 and the proof of the bicommutant theorem (Theorem 2.19) reveal that

$$B\left((\ell^2 G)^n\right)^\lambda = \left(\lambda(\mathbb{C}G)\,\mathrm{id}_{(\ell^2 G)^n}\right)' = M_n(\lambda(\mathbb{C}G)') = M_n(\mathcal{R}(G)).$$

So $B\left((\ell^2 G)^n\right)^\lambda$ is equivalently the weak closure of the $(n \times n)$-matrices $M_n(\mathbb{C}G)$ embedded as unital *-subalgebra of $B\left((\ell^2 G)^n\right)$ by matrix multiplication from the right using ρ. The *-operation acts on $M_n(\mathbb{C}G)$ by transposition and involuting the entries as in Example 2.22.

Example 2.29 A von Neumann algebra M whose center $Z(M) = M \cap M'$ equals $\mathbb{C}\mathrm{id}_H$ is called a *factor*. It is not hard to see that for the free group F_n on $n \geq 2$ letters, the group von Neumann algebra $\mathcal{R}(F_n)$ is a factor. In fact, $\mathcal{R}(G)$ is a factor if and only if G is *i.c.c.* meaning that every nontrivial conjugacy class in G is infinite. Here is an open problem in von Neumann algebras.

Question 2.30 (Free Factor Problem) Let $n > m \geq 2$. Are $\mathcal{R}(F_n)$ and $\mathcal{R}(F_m)$ isomorphic as von Neumann algebras?

To make this question meaningful, we need to define what morphisms of von Neumann algebras should be. This is somewhat subtle and following [138], it is best done once an intrinsic definition of von Neumann algebras is available: Without reference to some Hilbert space, Sakai [154] characterized a von Neumann algebra as an abstract C^*-algebra M that admits a *predual*: a Banach space V whose *dual Banach space* $V^* = B(V, \mathbb{C})$, defined as below Theorem 2.17, is (isometrically) isomorphic to M. Preduals are unique in the strong sense that for any two isomorphisms $m_V \colon V^* \to M$ and $m_W \colon W^* \to M$, there exists a unique isomorphism $F \colon W \to V$ with $m_V = m_W \circ F^*$. Now a *morphism* of von Neumann algebras is a C^*-homomorphism $f \colon M \to N$ which admits a *predual*: there exist isomorphisms $m \colon V^* \to M$ and $n \colon W^* \to N$ and a Banach space morphism

$F: W \to V$ such that $n \circ F^* = f \circ m$. Isomorphisms of von Neumann algebras are invertible morphisms.

The *weak-* topology* on the dual Banach space V^* is the coarsest topology that makes the evaluation maps $E_x: V^* \to \mathbb{C}$, $\varphi \mapsto \varphi(x)$ for all $x \in V$ continuous. By uniqueness, the weak-* topologies of all preduals induce one and the same topology on a von Neumann algebra M. It is called the *ultraweak operator topology* of M. As it turns out, a C^*-morphism $M \to N$ of von Neumann algebras is a morphism of von Neumann algebras if and only if it is ultraweakly-ultraweakly continuous. Even better, an abstract *-isomorphism $M \to N$ is automatically an ultraweak-ultraweak homeomorphism [20, III.2.2.12]. So long story short: in Question 2.30, we could have equivalently asked whether $\mathcal{R}(F_n)$ and $\mathcal{R}(F_m)$ are isomorphic as *-algebras.

Of course neither are F_n and F_m isomorphic as groups nor are $\mathbb{C}F_n$ and $\mathbb{C}F_m$ isomorphic as $\mathbb{C}$-algebras. An abstract formulation of the easy reason is that the functor $\mathbb{C}[\cdot]$ from groups to $\mathbb{C}$-algebras is left adjoint to the unit group functor $(\cdot)^\times$ from $\mathbb{C}$-algebras to groups, so an algebra homomorphism $\mathbb{C}F_n \to \mathbb{C}$ is specified by exactly n elements in $\mathbb{C}^\times$. But those homomorphisms do not extend to $\mathcal{R}(F_n)$ and it is notoriously hard to keep track of what is happening after taking weak closures. For more on the theory of factors, we recommend Jones' lecture notes [83].

Exercises

2.2.1 Work out the details in the proof of the Riesz lemma (Theorem 2.18).

2.2.2 Let V be a normed space with completion $\overline{V}$ and let W be a complete normed space. Show that a bounded linear operator $T: V \to W$ extends uniquely to a bounded linear operator $\overline{T}: \overline{V} \to W$ and $\|\overline{T}\| = \|T\|$.

2.2.3 Let H be a separable Hilbert space. We consider the three topologies τ_{weak}, τ_{strong} and τ_{norm} on $B(H)$.

  (i) We have $\tau_{\text{weak}} \subset \tau_{\text{strong}} \subset \tau_{\text{norm}}$.
 (ii) If H is infinite dimensional, then both of these inclusions are proper. *Hint: Fix $H \cong \ell^2$ and consider operators which delete or shift members of sequences.*
(iii) The involution $T \mapsto T^*$ on $B(H)$ is weakly and norm continuous but not strongly continuous unless H is finite dimensional.

2.2.4 Let $T, U \in B(H)$.

  (i) If U is a partial isometry, so is U^*.
 (ii) The operator T is self-adjoint if and only if $\langle Tx, x \rangle \in \mathbb{R}$ for all $x \in H$. *Hint: Can one compute $\langle Tx, y \rangle$ for all $x, y \in H$ if one only knows the values $\langle Tx, x \rangle$ for $x \in H$?*
(iii) If T is self-adjoint, we have $\|T\| = \sup_{\|x\|=1} |\langle Tx, x \rangle|$.

2.2.5 *In this exercise we construct the* polar decomposition *of an operator* $T \in B(H, K)$.

(i) Let $A \in B(H)$ be positive. Show that there is a unique positive operator $B \in B(H)$ such that $B^2 = A$. *Hint: You may use that the power series about zero of the function $f(z) = \sqrt{1 - z}$ converges absolutely for $|z| \leq 1$.*

(ii) Show that there exist a partial isometry $U \in B(H, K)$ and a positive operator $P \in B(H)$ such that $T = UP$. *Hint: Set $P^2 = T^*T$. What effect should U have on $\operatorname{im} P$? And on $\ker P$?*

(iii) Show that U and P can be arranged to satisfy $\ker U = \ker P$ and that requiring this determines them uniquely. We set $|T| = P$ and call $T = U|T|$ the *right-handed polar decomposition* of T.

(iv) Construct a *left-handed polar decomposition* $T = |T|U$ by a careful use of adjoints. What condition makes it unique?

(v) Show that if $H = K$ and T is *normal* (commutes with T^*), then the partial isometries and the positive operators in the right- and left-handed polar decompositions agree (and commute).

2.2.6 Let $T \in B(H)$.

(i) The operator T is invertible if and only if T has dense image and T is additionally *bounded from below* meaning $\|Tx\| \geq \varepsilon\|x\|$ for some $\varepsilon > 0$ and all $x \in H$.

(ii) If T and T^* are bounded from below, then T is invertible.

(iii) Every $T \in B(H)$ is a linear combination of two self-adjoints.

(iv) Every $T \in B(H)$ is a linear combination of four unitaries. *Hint: Review Exercise 2.2.5 (i) from above.*

2.2.7 Let $\mathcal{A} \subseteq B(H)$ be a C^*-algebra and let $T \in \mathcal{A}$ be invertible in $\mathcal{A}$. Show that the partial isometry and the positive operator in (both) the polar decomposition(s) of T lie in $\mathcal{A}$. *Remark: If $\mathcal{A}$ is even a von Neumann algebra, the conclusion holds true without assuming T was invertible.*

2.2.8 Let H be a Hilbert space, let $M \subset B(H)$ be a von Neumann algebra and let $\Omega \in H$ be a vector. We say that Ω is *cyclic* for M if $M\Omega \subset H$ is dense. We say that Ω is *separating* for M if for $T \in M$ we have $T\Omega = 0$ if and only if $T = 0$.

(i) Show that Ω is cyclic for M if and only if Ω is separating for M'.

(ii) Show that the unit element $e \in G$ constitutes a cyclic and separating vector $e \in \ell^2 G$ for the group von Neumann algebra $\mathcal{R}(G) = \rho(\mathbb{C}G)''$.

2.2.9 Consider the conjugate linear involution $J : \ell^2 G \to \ell^2 G$ given by $\sum_g c_g g \mapsto \sum_g \overline{c_g} g^{-1}$.

(i) Show that $J(Te) = T^*e$ for all $T \in \mathcal{R}(G) = \rho(\mathbb{C}G)''$. *Hint: First consider $T \in \rho(\mathbb{C}G) \subseteq \mathcal{R}(G)$. Remember that the adjoint map is not strongly continuous.*

(ii) Show that $\langle Jx, Jy \rangle = \langle y, x \rangle$ and $JTJ(Se) = ST^*(e)$ where $x, y \in \ell^2 G$ and $S, T \in \mathcal{R}(G)$.

(iii) Show that $JR(G)J \subseteq R(G)'$. *Hint: Use* (ii) *above and Exercise* 2.2.8 (ii).
(iv) Show that the formula $J(Te) = T^*e$ also holds for $T \in R(G)'$. *Hint: It is enough to show that $J(Te)$ and T^*e are mapped to the same complex number under $\langle \cdot, Ae \rangle$ for all $A \in R(G)$.*
(v) Show that $JR(G)J = R(G)'$. *Hint: Use Exercise* 2.2.8 (i) *and* 2.2.8 (ii).
(vi) Conclude $\rho(\mathbb{C}G)' = \lambda(\mathbb{C}G)''$. In words: left and right group von Neumann algebras are commutants of one another.

2.3   Trace and Dimension

Here is what makes group von Neumann algebras so useful.

Definition 2.31 Let $e \in \ell^2 G$ be the canonical basis vector given by the unit element in G. The linear functional

$$\mathrm{tr}_{R(G)} : R(G) \to \mathbb{C}, \quad T \mapsto \langle e, Te \rangle$$

is called the *von Neumann trace* or simply the *trace* of $R(G)$.

Of course, a linear functional only deserves to be called "trace" if it satisfies the *trace property* $\mathrm{tr}_{R(G)}(ST) = \mathrm{tr}_{R(G)}(TS)$ for all $S, T \in R(G)$. This can be checked by an easy calculation if $S, T \in \mathbb{C}G \subset R(G)$ and thus holds for all of $R(G)$ because $\mathrm{tr}_{R(G)}$ is weakly continuous by definition. To make the reader value the availability of a trace in $R(G)$ from the very start, we should say that for an infinite dimensional Hilbert space H there does not exist any nonzero linear functional $\mathrm{tr} : B(H) \to \mathbb{C}$ satisfying $\mathrm{tr}(ST) = \mathrm{tr}(TS)$, not even if we do not impose any continuity requirement whatsoever. The von Neumann trace does extend, however, to the amplified group von Neumann algebra of Example 2.28 by summing up the traces of the diagonal entries which clearly maintains the trace property. The survival of the trace when passing from linear algebra to the infinite-dimensional setting of group von Neumann algebras will later in the course allow us to recover two further notions from linear algebra: dimension and determinant.

Example 2.32 To keep up with our running example we spell out that the trace of $f \in L^\infty[-\pi, \pi] \cong R(\mathbb{Z})$ is given by $\mathrm{tr}_{R(\mathbb{Z})}(f) = \frac{1}{2\pi} \int_{-\pi}^{\pi} f(x)\mathrm{d}x$. For any measurable subset $A \subset [-\pi, \pi]$ the characteristic function $\chi_A \in L^\infty[-\pi, \pi]$ satisfies $\chi_A = \overline{\chi_A} = \chi_A^2$ and thus is a projection. We have $\mathrm{tr}_{R(\mathbb{Z})}(\chi_A) = \frac{\lambda(A)}{2\pi}$ where λ denotes Lebesgue measure. So the trace of a projection in $M_n(R(\mathbb{Z}))$ can take any real number in $[0, n]$ as value. Note that projections in $M_n(\mathbb{C})$ must have integer traces in $[0, n]$.

We would like to define traces also of endomorphisms (bounded G-operators) of an abstract Hilbert space H with isometric linear left G-action. Say H comes equipped with a fixed isometric linear embedding $i : H \hookrightarrow (\ell^2 G)^n$ for some n which is equivariant with respect to the diagonal λ-action on $(\ell^2 G)^n$. Then H is identified with a closed G-invariant subspace of $(\ell^2 G)^n$ and we know what to do. Let $\mathrm{pr}_{i(H)} \in B((\ell^2 G)^n)^\lambda$ be the orthogonal projection onto $i(H)$ and let $B(H)^G$ denote the endomorphism set of H.

Proposition 2.33 *The linear functional* $B(H)^G \to \mathbb{C}$ *given by*

$$T \mapsto \mathrm{tr}_{\mathcal{R}(G)}(i \circ T \circ \mathrm{pr}_{i(H)})$$

is independent of the choice of embedding $i \colon H \hookrightarrow (\ell^2 G)^n$.

Proof Let $j \colon H \hookrightarrow (\ell^2 G)^m$ be another linear isometric G-embedding. It is clear that the trace is invariant under stabilization

$$B((\ell^2 G)^n)^\lambda \to B((\ell^2 G)^{n+k})^\lambda, \quad T \mapsto T \oplus 0$$

so that we may assume $n = m$. Using the inverse mapping theorem (Theorem 2.16) the two embeddings i and j define a unitary $j \circ i^{-1} \colon i(H) \to j(H)$ which we extend to a partial isometry u on $(\ell^2 G)^n$ by setting it equal to zero on $i(H)^\perp$. By construction, u satisfies $j = u \circ i$ and hence, taking adjoints, $\mathrm{pr}_{j(H)} = \mathrm{pr}_{i(H)} \circ u^*$. Thus

$$\mathrm{tr}_{\mathcal{R}(G)}(j \circ T \circ \mathrm{pr}_{j(H)}) = \mathop{\mathrm{tr}}_{\mathcal{R}(G)}(u \circ i \circ T \circ \mathrm{pr}_{i(H)} \circ u^*) =$$

$$= \mathrm{tr}_{\mathcal{R}(G)}(i \circ T \circ \mathrm{pr}_{i(H)} \circ u^* u) =$$

$$= \mathop{\mathrm{tr}}_{\mathcal{R}(G)}(i \circ T \circ \mathrm{pr}_{i(H)} \circ \mathrm{pr}_{i(H)}) =$$

$$= \mathrm{tr}_{\mathcal{R}(G)}(i \circ T \circ \mathrm{pr}_{i(H)}). \qquad \square$$

So to obtain a well-defined trace we do not need a particular embedding $H \hookrightarrow (\ell^2 G)^n$, only existence matters. This explains the following definition.

Definition 2.34 A *Hilbert $\mathcal{L}(G)$-module* (also *Hilbert G-module* or just *Hilbert module*) is a Hilbert space H with linear isometric left G-action such that there exists a linear isometric G-embedding $H \hookrightarrow (\ell^2 G)^n$ for some n.

To justify the terminology "$\mathcal{L}(G)$-module", one observes that the G-action on H extends uniquely to a linear $\mathcal{L}(G)$-action as follows. We write $u \in \mathcal{L}(G)$ as a strong limit of group ring elements $u = \text{s-}\lim_{i \in I} \lambda(\sum_{g \in G} c_{i,g} g)$ and set $u \cdot x = \lim_{i \in I}(\sum_{g \in G} c_{i,g} g \cdot x)$ for $x \in H$. Well-definedness and uniqueness is easily established with the help of any G-embedding $H \hookrightarrow (\ell^2 G)^n$.

More precisely, a Hilbert module in the above sense should be called a *finitely generated* Hilbert module as opposed to a *general* Hilbert module for which we would require existence of a linear, isometric G-embedding into $\ell^2 G \otimes K$ for some possibly infinite dimensional Hilbert space K. Here $\ell^2 G \otimes K$ is the Hilbert space tensor product as discussed in (2.1) on p. 12, with the left G-action defined on elementary tensors by $h(g \otimes x) = (hg) \otimes x$. General Hilbert modules will not pop up before Chap. 4 so that for now we will leave the assumption of finite generation implicit.

We will say that a Hilbert module H is *free* if a G-equivariant unitary $H \xrightarrow{\sim} (\ell^2 G)^n$ can be chosen as embedding. Morphisms of Hilbert $\mathcal{L}(G)$-modules are bounded G-operators. Proposition 2.33 tells us that endomorphisms of Hilbert $\mathcal{L}(G)$-modules have a canonical trace which we still denote by $\mathrm{tr}_{\mathcal{R}(G)}$. Let us analyze what properties this trace has on offer.

Theorem 2.35 (von Neumann Trace) *Let H be a Hilbert $\mathcal{L}(G)$-module.*

- *(i)* Linearity. *The trace $\mathrm{tr}_{\mathcal{R}(G)}$ is $\mathbb{C}$-linear.*
- *(ii)* Weak continuity. *The trace $\mathrm{tr}_{\mathcal{R}(G)}$ is weakly continuous.*
- *(iii)* Trace property. *Let $s, t \in B(H)^G$. then $\mathrm{tr}_{\mathcal{R}(G)}(st) = \mathrm{tr}_{\mathcal{R}(G)}(ts)$.*
- *(iv)* Faithfulness. *Let $t \in B(H)^G$. Then $\mathrm{tr}_{\mathcal{R}(G)}(t^*t) = 0$ if and only if $t = 0$.*
- *(v)* Positivity. *Let $s, t \in B(H)^G$ and $s \leq t$. Then $\mathrm{tr}_{\mathcal{R}(G)}(s) \leq \mathrm{tr}_{\mathcal{R}(G)}(t)$.*

Proof Fix an embedding $i\colon H \to (\ell^2 G)^n$. Linearity is clear. To see (ii) we only have to convince ourselves that the map $B(H)^G \to B((\ell^2 G)^n)^\lambda$, $t \mapsto \bar{t}$, given by $\bar{t} = i \circ t \circ \mathrm{pr}_{i(H)}$, is weakly continuous. So let $t \in B(H)^G$ be weak limit of the net $(t_j)_{j \in I}$ in $B(H)^G$. Then for all $x, y \in (\ell^2 G)^n$ we have

$$\lim_{j \in I} \langle x, \bar{t_j} y \rangle = \lim_{j \in I} \langle x, i \circ t_j \circ \mathrm{pr}_{i(H)}(y) \rangle = \lim_{j \in I} \langle \mathrm{pr}_{i(H)}(x), t_j(\mathrm{pr}_{i(H)}(y)) \rangle =$$

$$= \langle \mathrm{pr}_{i(H)}(x), t\, \mathrm{pr}_{i(H)} y \rangle = \langle x, \bar{t} y \rangle,$$

thus $\bar{t_i} \to \bar{t}$ weakly.

Since $\mathrm{pr}_{i(H)} \circ i = \mathrm{id}_H$, we have $\overline{st} = \bar{s}\,\bar{t}$. Therefore (iii) follows from the trace property in the amplified algebra $B((\ell^2 G)^n)^G$.

To show (iv) we first note that since $i^* = \mathrm{pr}_{i(H)}$, we have moreover $\overline{t^*} = \bar{t}^*$. For $\overline{t^*t} \in M_n(\mathcal{R}(G))$ we thus have

$$\mathrm{tr}_{\mathcal{R}(G)}(\overline{t^*t}) = \sum_{j=1}^{n} \langle e, (\overline{t^*t})_{jj} e \rangle = \sum_{j=1}^{n} \langle e, (\bar{t}^*\bar{t})_{jj} e \rangle = \sum_{j=1}^{n} \sum_{k=1}^{n} \langle e, (\bar{t}^*)_{jk} \bar{t}_{kj} e \rangle =$$

$$= \sum_{j=1}^{n} \sum_{k=1}^{n} \langle e, (\bar{t}_{kj})^* \bar{t}_{kj} e \rangle = \sum_{j,k=1}^{n} \| \bar{t}_{kj} e \|^2$$

which in case $\mathrm{tr}_{\mathcal{R}(G)}(\overline{t^*t}) = 0$ gives $\bar{t}_{jk} e = 0$ for all $1 \leq j, k \leq n$. For any other basis vector $g \in \ell^2(G)$ we then likewise have $\bar{t}_{jk} g = \bar{t}_{jk}(ge) = g\bar{t}_{jk} e = 0$. Thus $\mathrm{tr}_{\mathcal{R}(G)}(t^*t) = 0$ implies $t = 0$.

To see (v) we only have to show $\mathrm{tr}_{\mathcal{R}(G)}(r) \geq 0$ if $r \in B(H)^G$ is positive. But this holds by definition because $\bar{r}$ is likewise positive. □

In Sect. 2.1 we listed various ways in which a new Hilbert space can arise out of old Hilbert spaces. All of these constructions extend from Hilbert spaces to Hilbert $\mathcal{L}(G)$-modules.

(i) If H is a Hilbert G-module, then so is every closed G-invariant subspaces $K \subset H$. Just restrict any embedding $H \hookrightarrow (\ell^2 G)^n$ to K. We call K a *Hilbert submodule*.

(ii) If $K \subset H$ is a closed G-invariant subspace of a Hilbert G-module H, then the quotient H/K is a Hilbert module since it can be identified with the closed G-invariant subspace $K^\perp \subset H$. We call H/K a *Hilbert quotient module* or *Hilbert factor module*.

(iii) For Hilbert G-modules H_1 and H_2 with embeddings $i_1 \colon H_1 \hookrightarrow (\ell^2 G)^{n_1}$ and $i_2 \colon H_2 \hookrightarrow (\ell^2 G)^{n_2}$ we obtain an embedding $i_1 \oplus i_2 \colon H_1 \oplus H_2 \hookrightarrow (\ell^2 G)^{n_1 + n_2}$ showing that the *Hilbert direct sum* $H_1 \oplus H_2$ is a Hilbert module.

(iv) Let H_1 be a Hilbert G_1-module and let H_2 be a Hilbert G_2-module. Pick embeddings $i_1 \colon H_1 \hookrightarrow (\ell^2 G_1)^n$ and $i_2 \colon H_2 \hookrightarrow (\ell^2 G_2)^m$. These tensor up to give an embedding

$$i = i_1 \otimes i_2 \colon H_1 \otimes H_2 \hookrightarrow (\ell^2 G_1)^n \otimes (\ell^2 G_2)^m \cong (\ell^2 G_1 \otimes \ell^2 G_2)^{nm}$$

$$\cong (\ell^2 (G_1 \times G_2))^{nm}$$

where all isomorphism are canonical, see the discussion in (2.1) on page 12. This shows that $H_1 \otimes H_2$ is a Hilbert $(G_1 \times G_2)$-module called the *Hilbert tensor product* of H_1 and H_2

(v) Here is a simple but important new concept. Let $G_0 \le G$ be a subgroup of finite index m. We have a functor $\mathrm{res}^G_{G_0}$ from Hilbert G-modules to Hilbert G_0-modules obtained by restricting the group action from G to G_0. For the embedding it is enough to observe that a system of representatives for G/G_0 determines a G_0-unitary $\mathrm{res}^G_{G_0} \ell^2(G) \cong (\ell^2 G_0)^m$. We call $\mathrm{res}^G_{G_0}(H)$ a *restricted Hilbert module*.

The von Neumann trace behaves well with respect to constructing new Hilbert modules out of old Hilbert modules. More precisely, if a Hilbert module arises as a direct product, tensor product or by restriction, the new traces can be expressed in terms of the old ones as follows.

Theorem 2.36 (Computing von Neumann Traces)

(i) Additivity. *Suppose we have a commutative diagram of Hilbert modules*

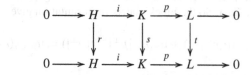

with exact rows. Then

$$\mathrm{tr}_{\mathcal{R}(G)}(s) = \mathrm{tr}_{\mathcal{R}(G)}(r) + \mathrm{tr}_{\mathcal{R}(G)}(t).$$

(ii) Multiplicativity. *Let H_1 be a Hilbert G_1-module and let H_2 be a Hilbert G_2-module. Two morphisms $s \in B(H_1)^{G_1}$ and $t \in B(H_2)^{G_2}$ define a morphism $s \otimes t \in B(H_1 \otimes H_2)^{G_1 \times G_2}$ such that*

$$\mathrm{tr}_{\mathcal{R}(G_1 \times G_2)}(s \otimes t) = \mathrm{tr}_{\mathcal{R}(G_1)}(s) \cdot \mathrm{tr}_{\mathcal{R}(G_2)}(t).$$

(iii) Restriction. *Let H be a Hilbert G-module and let $G_0 \leq G$ be a subgroup of finite index. Then for every $s \in B(H)^G$ we have*

$$\mathrm{tr}_{\mathcal{R}(G_0)}(\mathrm{res}_{G_0}^G(s)) = [G : G_0]\, \mathrm{tr}_{\mathcal{R}(G)}(s).$$

Proof We start with (i). By the inverse mapping theorem (Theorem 2.16) both $i : H \to \ker p$ and $p^{\perp} = p_{|(\ker p)^{\perp}} : (\ker p)^{\perp} \to L$ are invertible and the latter means that every short exact sequence of Hilbert modules is split. Let $i = u_1|i|$ and $p^{\perp} = |p^{\perp}|u_2$ be polar decompositions of i and $p^{\perp}$, respectively. Arguing as in Exercise 2.2.7 we see that all appearing operators are themselves G-equivariant. We obtain a unitary $u = u_1 \oplus u_2^* : H \oplus L \to K$ and a commutative diagram

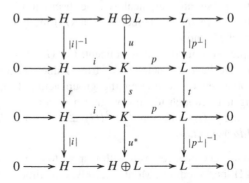

where the top and the bottom row are the standard short exact sequence. The diagram tells us that the endomorphism u^*su is given in block form as $\begin{pmatrix} |i|r|i|^{-1} & * \\ 0 & |p^{\perp}|^{-1}t|p^{\perp}| \end{pmatrix}$. Let $j = j_1 \oplus j_2 : H \oplus L \hookrightarrow (\ell^2 G)^{n_1} \oplus (\ell^2 G)^{n_2}$ be an embedding coming from two embeddings j_1 and j_2 of H and L, respectively. Since u is unitary, $j \circ u^*$ is an (isometric) embedding $K \hookrightarrow (\ell^2 G)^n$. Thus $\mathrm{tr}_{\mathcal{R}(G)}(s) = \mathrm{tr}_{\mathcal{R}(G)}(u^*su)$. So the block matrix representation gives

$$\mathrm{tr}_{\mathcal{R}(G)}(s) = \mathrm{tr}_{\mathcal{R}(G)}(|i|r|i|^{-1}) + \mathrm{tr}_{\mathcal{R}(G)}(|p^{\perp}|^{-1}t|p^{\perp}|) = \mathrm{tr}_{\mathcal{R}(G)}(r) + \mathrm{tr}_{\mathcal{R}(G)}(t)$$

by the trace property, Theorem 2.35 (iii).

We now prove (ii). The morphisms $s \in B(H_1)^{G_1}$ and $t \in B(H_2)^{G_2}$ define the morphism $s \otimes t \in B(H_1 \otimes H_2)^{G_1 \times G_2}$ by requiring $(s \otimes t)(x_1 \otimes x_2) = s(x_1) \otimes t(x_2)$

on elementary tensors. Retaining the notation from the previous proof, we have a diagram

$$
\begin{CD}
0 @>>> H_1 \otimes H_2 @>i_1 \otimes i_2>> (\ell^2 G_1)^n \otimes (\ell^2 G_2)^m @>pr>> \operatorname{im}(i_1 \otimes i_2)^\perp @>>> 0 \\
@. @VV s \otimes t V @VV \bar{s} \otimes \bar{t} V @VV 0 V \\
0 @>>> H_1 \otimes H_2 @>i_1 \otimes i_2>> (\ell^2 G_1)^n \otimes (\ell^2 G_2)^m @>pr>> \operatorname{im}(i_1 \otimes i_2)^\perp @>>> 0.
\end{CD}
$$

By additivity (i) we obtain $\operatorname{tr}_{\mathcal{R}(G_1 \times G_2)}(s \otimes t) = \operatorname{tr}_{\mathcal{R}(G_1 \times G_2)}(\bar{s} \otimes \bar{t})$. The proof concludes with the computation

$$
\operatorname{tr}_{\mathcal{R}(G_1 \times G_2)}(\bar{s} \otimes \bar{t}) = \langle e \otimes e, (\bar{s} \otimes \bar{t})(e \otimes e) \rangle = \langle e \otimes e, \bar{s}(e) \otimes \bar{t}(e) \rangle =
$$
$$
= \langle e, \bar{s}(e) \rangle \cdot \langle e, \bar{t}(e) \rangle = \operatorname{tr}_{\mathcal{R}(G_1)}(s) \cdot \operatorname{tr}_{\mathcal{R}(G_2)}(t)
$$

where we applied the definition of inner product on $H_1 \otimes_{\mathbb{C}} H_2$ from page 12.

To see (iii) we choose a system of representatives $g_1, \ldots, g_m \in G$ for the cosets in G/G_0 giving rise to a unitary of Hilbert spaces $u \colon (\ell^2 G_0)^m \to \ell^2 G$ determined by $(h_1, \ldots, h_m) \mapsto h_1 g_1 + \cdots + h_m g_m$ for $h_i \in G_0$. This unitary is moreover G_0-equivariant when viewed as a map $u \colon (\ell^2 G_0)^m \to \operatorname{res}^G_{G_0}(\ell^2 G)$. Thus $u^* \colon \operatorname{res}^G_{G_0}(\ell^2 G) \to (\ell^2 G_0)^m$ is an embedding showing that $\operatorname{res}^G_{G_0}(\ell^2 G)$ is a Hilbert G_0-module. For any $s \in \mathcal{R}(G) = B(\ell^2 G)^\lambda$ we have $\overline{\operatorname{res}}^G_{G_0}(s) = u^* s u \in M_m(\mathcal{R}(G_0))$. A moment's thought gives

$$
\overline{\operatorname{res}}^G_{G_0}(s)_{ij}(e) = \operatorname{pr}_{\ell^2 G_0}(g_j s(e) g_i^{-1})
$$

from which it follows that

$$
\operatorname{tr}_{\mathcal{R}(G_0)} \overline{\operatorname{res}}^G_{G_0}(s)_{ii} = \langle e, g_i s(e) g_i^{-1} \rangle = \langle g_i^{-1} e g_i, s(e) \rangle = \langle e, s(e) \rangle = \operatorname{tr}_{\mathcal{R}(G)}(s)
$$

independent of i. Thus we have $\operatorname{tr}_{\mathcal{R}(G_0)}(\operatorname{res}^G_{G_0}(s)) = m \operatorname{tr}_{\mathcal{R}(G)}(s)$. The general case $s \in B(H)^G$ follows by composing a given embedding of H with the unitary $u^* \oplus \cdots \oplus u^*$ to obtain an embedding of $\operatorname{res}^G_{G_0}(H)$. Then one can view $\overline{\operatorname{res}}^G_{G_0}(s)$ as an element of $M_n(M_m(\mathcal{R}(G_0)))$ that satisfies the relation $\overline{\operatorname{res}}^G_{G_0}(s)_{ii} = \operatorname{res}^G_{G_0}(\bar{s}_{ii})$. Therefore $\operatorname{tr}_{\mathcal{R}(G_0)}(\operatorname{res}^G_{G_0}(s)) = m \operatorname{tr}_{\mathcal{R}(G)}(s)$ follows from the above. $\square$

Definition 2.37 Let H be a Hilbert $\mathcal{L}(G)$-module. We define the *von Neumann dimension* of H as

$$
\dim_{\mathcal{R}(G)}(H) = \operatorname{tr}_{\mathcal{R}(G)}(\operatorname{id}_H).
$$

Example 2.38 If G is the trivial group, a Hilbert module is just a finite dimensional inner product space. Since $\mathcal{R}(G) = \mathbb{C}$, we should have $\dim_{\mathcal{R}(G)} = \dim_{\mathbb{C}}$ and indeed, this notational collision is deliberate: von Neumann dimension with trivial G is ordinary complex vector space dimension. A computer scientist might say "we have overloaded the dimension function".

Example 2.39 If G is a finite group, the underlying complex vector space of a Hilbert G-module H is still of finite complex dimension. Setting G_0 equal to the trivial subgroup, the restriction property of the trace, Theorem 2.36 (iii), shows that $\dim_{\mathcal{R}(G)}(H) = \frac{\dim_{\mathbb{C}} H}{|G|}$.

Note that $\mathbb{C}^k$ with trivial G-action is a Hilbert G-module because we can embed it by $i \colon \mathbb{C}^k \hookrightarrow (\ell^2 G)^k$ sending the k-th coordinate z_k to $\frac{z_k}{\sqrt{n}}(g_1 + \cdots + g_n)$ in the k-th coordinate of $(\ell^2 G)^k$ for $G = \{e = g_1, g_2, \ldots, g_n\}$. The formula from above gives $\dim_{\mathcal{R}(G)}(\mathbb{C}^k) = \frac{k}{n}$.

Example 2.40 If $G = \mathbb{Z}$, then every measurable subset $A \subseteq [-\pi, \pi]$ gives rise to a closed subspace $L^2 A \subseteq L^2[-\pi, \pi] \cong \ell^2(\mathbb{Z})$. In Example 2.32 we saw that $\mathrm{tr}_{\mathcal{R}(\mathbb{Z})}\, \chi_A = \frac{\lambda(A)}{2\pi}$ so that $\dim_{\mathcal{R}(\mathbb{Z})}(L^2 A)^k = \frac{k\lambda(A)}{2\pi}$. This shows that von Neumann dimension can take any nonnegative real number as value.

Example 2.41 Let $H \leq G$ be a subgroup. The Hilbert space $\ell^2(G/H)$ has an isometric, linear G-action defined on the orthonormal basis G/H by left translation of cosets. We claim that this action turns $\ell^2(G/H)$ into a Hilbert $\mathcal{L}(G)$-module if and only if H is finite. In the latter case we obtain $\dim_{\mathcal{R}(G)} \ell^2(G/H) = \frac{1}{|H|}$.

Indeed, if $H = \{h_1 = e, h_2, \ldots, h_n\}$ is finite, then sending the element gH to $\frac{1}{\sqrt{n}}(gh_1 + \cdots + gh_n)$ defines a well-defined, linear, isometric G-embedding $i \colon \ell^2(G/H) \hookrightarrow \ell^2 G$. In this way $\ell^2(G/H)$ is embedded in $\ell^2 G$ as the subspace of all Fourier series with constant coefficient throughout left H-cosets. The projection onto this subspace is given by right multiplication with $\frac{1}{n}(h_1 + \cdots + h_n)$. The unit matrix coefficient of this projection is $\frac{1}{n}$ showing that $\dim_{\mathcal{R}(G)} \ell^2(G/H) = \frac{1}{|H|}$.

On the other hand $H \in \ell^2(G/H)$ always gives an H-invariant vector, whether H is finite or not. So if there is a linear G-embedding $\ell^2(G/H) \hookrightarrow (\ell^2 G)^n$, there exists a nonzero H-invariant vector in $(\ell^2 G)^n$. Since G acts diagonally on $(\ell^2 G)^n$, any nonzero coordinate of this H-invariant vector gives a nonzero H-invariant vector $x \in \ell^2 G$. Let $x = \sum_{g \in G} c_g g$ be the Fourier series of x. Then H-invariance says $hx = x$ which means

$$\sum_{g \in G} c_{h^{-1}g} g = \sum_{g \in G} c_g g$$

so that $c_{h^{-1}g} = c_g$ for all $g \in G$ and $h \in H$. Thus the Fourier coefficients of x are constant throughout right H-cosets in G. If H is infinite, ℓ^2-summability says they all vanish, contradicting that x is nonzero.

Before we translate properties of von Neumann trace to properties of von Neumann dimension, we must point the reader's attention to an important peculiarity of the category of Hilbert spaces and thus also of Hilbert $\mathcal{L}(G)$-modules: a monomorphism which is also an epimorphism may fail to be an isomorphism! The problem is that in the diagram of Hilbert modules

$$H \xrightarrow{\ s\ } K \underset{t_2}{\overset{t_1}{\rightrightarrows}} L$$

it is enough that s have *dense image* to conclude $t_1 = t_2$ from $t_1 \circ s = t_2 \circ s$.

Example 2.42 Consider the polynomial $p(z) = z - 1$ acting on $L^2(S^1)$. A function $f \in \ker(p(z))$ must have constant Fourier coefficients but then Parseval's identity (or simply ℓ^2 summability) requires that they all vanish, so $f = 0$. For the same reason the adjoint $p^*(z) = z^{-1} - 1$ is injective, thus $\overline{\operatorname{im} p(z)} = \ker(p^*(z))^{\perp} = \{0\}^{\perp} = L^2(S^1)$, so $p(z)$ has dense image. But the constant function $1 \in L^2(S^1)$ has no preimage because its Fourier coefficients would have to satisfy $c_k = c_0$ and $c_{-k} = c_{-1}$ for all $k \geq 1$ as well as $c_0 - c_{-1} = 1$. There is no square summable way to make that happen.

To do justice to this phenomenon we introduce the following terminology.

Definition 2.43 A sequence $H \xrightarrow{i} K \xrightarrow{p} L$ of Hilbert modules is called *weakly exact* at K if $\ker p = \overline{\operatorname{im} i}$. A morphism $s: H \to K$ is called a *weak isomorphism* if $0 \longrightarrow H \xrightarrow{s} K \longrightarrow 0$ is weakly exact.

With these notions at our disposal we can set up a handy tool box for computing von Neumann dimension.

Theorem 2.44 (Computing von Neumann Dimension)

(i) Normalization. *We have* $\dim_{\mathcal{R}(G)}(\ell^2 G) = 1$.

(ii) Faithfulness. *For a Hilbert $\mathcal{L}(G)$-module H we have* $\dim_{\mathcal{R}(G)}(H) = 0$ *if and only if H is trivial.*

(iii) Outer regularity. *Let $\{H_i\}_{i \in I}$ be a system of Hilbert submodules of a Hilbert $\mathcal{L}(G)$-module H directed by containment "$\supseteq$". Then*

$$\dim_{\mathcal{R}(G)} \bigcap_{i \in I} H_i = \inf_{i \in I} \dim_{\mathcal{R}(G)} H_i.$$

(iv) Inner regularity. *Let $\{H_i\}_{i \in I}$ be a system of Hilbert submodules of a Hilbert $\mathcal{L}(G)$-module H directed by inclusion "$\subseteq$". Then*

$$\dim_{\mathcal{R}(G)} \overline{\bigcup_{i \in I} H_i} = \sup_{i \in I} \dim_{\mathcal{R}(G)} H_i.$$

(v) Additivity. *Let* $0 \longrightarrow H \xrightarrow{i} K \xrightarrow{p} L \longrightarrow 0$ *be a weakly exact sequence of Hilbert $\mathcal{L}(G)$-modules. Then*

$$\dim_{\mathcal{R}(G)} K = \dim_{\mathcal{R}(G)} H + \dim_{\mathcal{R}(G)} L.$$

(vi) Multiplicativity. *Let H_1 be a Hilbert $\mathcal{L}(G_1)$-module and let H_2 be a Hilbert $\mathcal{L}(G_2)$-module. Then*

$$\dim_{\mathcal{R}(G_1 \times G_2)} H_1 \otimes H_2 = \dim_{\mathcal{R}(G_1)} H_1 \cdot \dim_{\mathcal{R}(G_2)} H_2.$$

(vii) Restriction. *Let H be a Hilbert $\mathcal{L}(G)$-module and let $G_0 \leq G$ be a subgroup of finite index. Then*

$$\dim_{\mathcal{R}(G_0)} \operatorname{res}^G_{G_0}(H) = [G : G_0] \dim_{\mathcal{R}(G)} H.$$

Proof Property (i) is clear. Properties (ii), (vi) and (vii) are immediate consequences of Theorems 2.35 and 2.36. Property (iii) follows from (iv) and (v) by considering the system $\{H_i^{\perp}\}_{i \in I}$. It remains to show (iv) and (v). For (iv), let $L = \bigcup_{i \in I} H_i$. Given $x \in H$ and $\varepsilon > 0$, there exists $i_0 \in I$ such that $\operatorname{pr}_L(x)$ lies in an open ε-ball around $\operatorname{pr}_{H_{i_0}}(x)$ because $\operatorname{pr}_{H_{i_0}}(x)$ is the point closest to x in H_{i_0} by Exercise 2.1.2. Thus $\|\operatorname{pr}_L(x) - \operatorname{pr}_{H_i}(x)\| < \varepsilon$ for all $i \geq i_0$, so the net $(\operatorname{pr}_{H_i})_{i \in I}$ converges strongly, hence weakly, to pr_L. Since the trace is weakly continuous by Theorem 2.35 (ii), this gives

$$\dim_{\mathcal{R}(G)} L = \operatorname{tr}_{\mathcal{R}(G)}(\operatorname{pr}_L) = \lim_{i \in I} \operatorname{tr}_{\mathcal{R}(G)}(\operatorname{pr}_{H_i}) = \sup_{i \in I} \dim_{\mathcal{R}(G)} H_i.$$

We should start the proof of (v) with the heads-up that unlike an exact sequence, a weakly exact sequence of Hilbert modules need not split. Nevertheless, $i \oplus p^* \colon H \oplus L \to K$ is a weak isomorphism. Even better, for the partial isometry u in the polar decomposition of $i \oplus p^*$ we have $uu^* = \operatorname{pr}_{\overline{\operatorname{im}(i \oplus p^*)}} = \operatorname{id}_K$. Thus u is unitary and therefore K is isomorphic to $H \oplus L$. Now (v) follows from additivity of the trace, Theorem 2.36 (i).   □

This last theorem shall conclude our quick trip through functional analysis. We will however come back to it in Chap. 5, Sect. 5.2, when we will explain functional calculus in various operator algebras and the spectral theorem. For now, we have collected enough material to return to our original objective: equivariant topology.

Exercises

2.3.1 Let G be a finite group of order n. For every prime divisor $p \mid n$ construct a projection $P \in \mathcal{R}(G)$ with $\operatorname{tr}_{\mathcal{R}(G)}(P) = \frac{1}{p}$.

2.3.2

(i) Show that the group ring $\mathbb{C}G$ is *directly finite*: if $ab = e$ for $a, b \in \mathbb{C}G$, then also $ba = e$. *Hint: Consider right multiplication by a and b as operators on $\ell^2 G$.*

(ii) Extend this result from $\mathbb{C}$ to any field F of characteristic zero. *Hint: Only finitely many elements of F occur as coefficients in a and b.*

Remark: If F is a field of positive characteristic, direct finiteness of FG is open (in general) and known as Kaplansky's direct finiteness conjecture.

Chapter 3
ℓ^2-Betti Numbers of CW Complexes

3.1 G-CW Complexes

Let us consider a space X with a left action of a (discrete, countable) group G by homeomorphisms. Say X carries in addition the structure of a CW complex so that X is filtered by skeleta X_n. As usual, when two structures are given on one object, we want to reconcile them by imposing some compatibility. Recall that an *open n-cell* $E \subset X$ is a connected component of $X_n \setminus X_{n-1}$. An *open cell* is an open n-cell for some n.

Definition 3.1 We say that G acts *cellularly* on the CW complex X if for each open cell $E \subset X$ and each $g \in G$

(i) the translated set gE is again an open cell in X,
(ii) if gE intersects E in a nonempty set, then g leaves E pointwise fixed.

By invariance of domain, condition (i) is equivalent to requiring that the translation map defined by g be *cellular* (respect the filtration by skeleta). Condition (ii) might look a little less natural. It says that the isometry group S_3 of the standard 2-simplex does not act cellularly with respect to the CW structure indicated on the left in Fig. 3.1. Just observe that every $g \in S_3$ translates the only 2-cell to itself but g does not leave it fixed pointwise unless $g = e$; so condition (ii) is violated in the strongest sense. The group S_3 does act cellularly, though, after one barycentric subdivision as depicted on the right of Fig. 3.1.

This example illustrates the idea of condition (ii). It ensures the cellular triangulation is sufficiently fine to describe the group action in combinatorial terms. This is what we will explain next. Since X is a CW complex, we can choose pushout diagrams that provide us with a homeomorphism

$$X_n \setminus X_{n-1} \cong \coprod_{j \in J_n} \mathring{D}^n$$

© Springer Nature Switzerland AG 2019
H. Kammeyer, *Introduction to ℓ^2-invariants*, Lecture Notes in Mathematics 2247,
https://doi.org/10.1007/978-3-030-28297-4_3

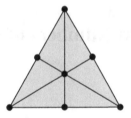

Fig. 3.1 Two different CW structures of the 2-simplex

where $\mathring{D}^n$ is the open n-disk. Condition (i) says G permutes the components of this space. So G acts on the index set J_n. Let $I_n = G\backslash J_n$ be the set of orbits and fix one element in each orbit $i \in I_n$. If H_i denotes the corresponding stabilizer group, we obtain the description $J_n = \coprod_{i \in I_n} G/H_i$, hence

$$X_n \setminus X_{n-1} \cong \coprod_{i \in I_n} G/H_i \times \mathring{D}^n$$

because G, thus G/H_i, is discrete. Condition (ii) implies that this map becomes a G-homeomorphism when G acts diagonally on $G/H_i \times \mathring{D}^n$ by left translation on G/H_i and trivially on $\mathring{D}^n$. This gives us the idea that just like a CW complex is obtained inductively by gluing in cells, it should be possible to construct a CW complex with cellular G-action by gluing in *equivariant G-cells* of the form $G/H \times D^n$ for some subgroup $H \leq G$.

Theorem 3.2 *Let X be a CW complex endowed with a left action by a discrete group G. The following are equivalent.*

(i) The group G acts cellularly on X.

(ii) The skeleta X_n are G-invariant subspaces and there exist pushouts in the category of G-spaces and G-maps

$$
\begin{array}{ccc}
\coprod_{i \in I_n} G/H_i \times S^{n-1} & \xrightarrow{\ q_n\ } & X_{n-1} \\
\downarrow{\scriptstyle i_n} & & \downarrow{\scriptstyle j_n} \\
\coprod_{i \in I_n} G/H_i \times D^n & \xrightarrow{\ Q_n\ } & X_n.
\end{array}
$$

Proof (ii) $\Rightarrow$ (i) is clear. If (i) holds, the above construction gives us diagrams as in (ii). These are diagrams in the category of G-spaces, as required, but we only know that they are pushouts in the category of spaces. To see that they are pushouts in the category of G-spaces, let $f \colon X_{n-1} \to Z$ and $g \colon \coprod_{i \in I_n} G/H_i \times D^n \to Z$ be G-maps with $f \circ q_n = g \circ i_n$. We obtain a unique map $u \colon X_n \to Z$ with $u \circ j_n = f$ and $u \circ Q_n = g$. It remains to show that u is G-equivariant. But X_n is the disjoint union of

X_{n-1} and $Q_n \left(\coprod_{i \in I_n} G/H_i \times \mathring{D}^n \right)$ and on these two subspaces the map u restricts to the maps f and g along the inclusions given by j_n and by the restriction of Q_n to $\coprod_{i \in I_n} G/H_i \times \mathring{D}^n$, respectively. Since f and g are G-equivariant by assumption, we obtain (ii). □

Definition 3.3 A *G-CW complex* is a CW complex X with an action by a discrete group G which satisfies either of the two conditions in Theorem 3.2. A G-CW complex X is called

- *finite type* if it has finitely many equivariant n-cells for every n,
- *finite* if it has finitely many equivariant cells altogether,
- *proper* if all stabilizer groups are finite,
- *free* if all stabilizer groups are trivial.

We remark that if G is not a discrete group but any locally compact Hausdorff group, we can still take Theorem 3.2 (ii) as the definition of a G-CW complex where we require that the G-action be continuous and all stabilizer groups H_i be closed subgroups. But again, unless otherwise stated, G will denote a discrete, countable group in what follows. Be aware that if G is infinite, a finite G-CW complex X must not be a compact space. In fact a G-CW complex is finite if and only if it is *cocompact* meaning that the quotient space $G \backslash X$ is compact.

The quotient space $N \backslash X$ of a G-CW complex by a normal subgroup $N \trianglelefteq G$ is a G/N-CW complex in a canonical way. For $N = G$ this says that the quotient space $G \backslash X$ is an ordinary CW-complex. If a group G acts cellularly on a CW complex X, then so does every subgroup $G_0 \leq G$. Therefore restricting the group action defines a functor $\operatorname{res}_{G_0}^G$ from G-CW complexes to G_0-CW complexes. Every equivariant G-cell $G/H \times D^n$ of X gives $[G : G_0]$-many equivariant G_0-cells $G_0 / H \cap G_0 \times D^n$ in $\operatorname{res}_{G_0}^G X$.

The subdivided 2-simplex from Fig. 3.1 is an example of a finite, proper S_3-CW complex which is not free. Figure 3.2 displays its equivariant cells. Here is a large supply of examples of free G-CW complexes.

Example 3.4 Let X be a connected, finite type CW complex. Then every Galois covering of X is a connected, finite type, free G-CW complex where G denotes the deck transformation group. In this case, the examples are exhaustive: for every

Fig. 3.2 S_3-equivariant cells of the subdivided 2-simplex: there are three 0-cells (circles, triangles, square), three 1-cells (dashed, dotted, solid) and one 2-cell (gray)

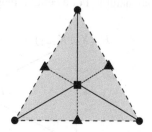

connected, finite type, free G-CW complex X the projection map $X \to G\backslash X$ is a Galois covering.

Exercises

3.1.1 Let G be a group and let $G_0 \leq G$ be a subgroup of finite index. Find the left adjoint to $\mathrm{res}_{G_0}^G : \underline{\mathcal{R}(G)\text{-mod}} \to \underline{\mathcal{R}(G_0)\text{-mod}}$.

3.1.2 Let X be a G-CW complex and let $G_0 \leq G$ be any subgroup. We obtain a G_0-CW complex $\mathrm{res}_{G_0}^G X$ by restricting the group action from G to G_0. This construction clearly gives a functor $\mathrm{res}_{G_0}^G : \underline{G\text{-CW}} \to \underline{G_0\text{-CW}}$. Find its left adjoint. *Hint: Quite a few details need attention in the construction. In this exercise (and only in this one!) a sketchy proof shall do.*

3.1.3 Consider the real line $\mathbb{R}$. We turn it into a CW complex X by decreeing that each integer is a 0-cell and that the intervals connecting adjacent integers are 1-cells. Let $t, r : \mathbb{R} \to \mathbb{R}$ be the transformations $t : x \mapsto x + 1$ and $r : x \mapsto -x$. Let $D_\infty = \langle t, r \rangle$ be the subgroup of the isometry group of $\mathbb{R}$ generated by t and r. By construction, this group comes with an action $D_\infty \curvearrowright X$.

(i) Make yourself aware that this action is <u>not</u> cellular.
(ii) Find a subdivided CW structure Y for which D_∞ does act cellularly. Show that the D_∞-CW complex Y is finite and proper but not free.

3.2   The ℓ^2-Completion of the Cellular Chain Complex

Let X be a G-CW complex. The translation map of every $g \in G$ defines self-homeomorphisms $(X_n, X_{n-1}) \xrightarrow{\sim} (X_n, X_{n-1})$ and thus an automorphism of the relative singular homology group $H_n(X_n, X_{n-1})$. The latter is by definition the n-th cellular chain group of X, so we see that the cellular chain complex $C_*(X)$ consists of left $\mathbb{Z}G$-modules. The differentials in $C_*(X)$ are the boundary maps in the long exact sequence of the triple (X_n, X_{n-1}, X_{n-2}). It is crucial in our context that these are natural: For a cellular map $f : X \to Y$ of CW complexes X and Y, the diagram

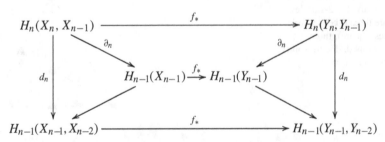

of singular homology groups commutes: the triangles commute by definition and the squares by naturality of singular pair sequences. Hence the outer square commutes and shows that the cellular chain boundaries are natural transformations $d_* : C_* \to C_{*-1}$. Specializing f to the cellular automorphism of X given by translation with $g \in G$, this implies that the boundary maps in the cellular chain complex $C_*(X)$ are G-equivariant. We thus have proven that the cellular chain complex $C_*(X)$ of a G-CW complex X is canonically a chain complex of left $\mathbb{Z}G$-modules. Of course, the chain map induced by a G-equivariant, cellular map of G-CW complexes consists of $\mathbb{Z}G$-homomorphisms. So the following proposition summarizes the discussion thus far.

Proposition 3.5 *The cellular chain complex defines a functor (C_*, d_*) from G-CW complexes to chain complexes of left $\mathbb{Z}G$-modules.*

An explicit description of the chain complex $C_*(X)$ becomes available after choosing pushout diagrams

$$
\begin{array}{ccc}
\coprod_{i \in I_n} G/H_i \times S^{n-1} & \xrightarrow{q_n} & X_{n-1} \\
\downarrow{i_n} & & \downarrow{j_n} \\
\coprod_{i \in I_n} G/H_i \times D^n & \xrightarrow{Q_n} & X_n
\end{array}
$$

whose existence is granted by Theorem 3.2 (ii). The arrow i_n is an inclusion as neighborhood deformation retract (a "cofibration") and hence the Mayer–Vietoris theorem for pushouts [85, Theorem 5.15, p. 56] says that (Q_n, q_n) induces an isomorphism

$$
H_n\left(\coprod_{i \in I_n} G/H_i \times D^n, \coprod_{i \in I_n} G/H_i \times S^{n-1}\right) \cong H_n(X_n, X_{n-1}).
$$

Since G is discrete, this gives

$$
C_n(X) \cong \bigoplus_{i \in I_n} \bigoplus_{G/H_i} H_n(D^n, S^{n-1}) \cong \bigoplus_{i \in I_n} \mathbb{Z}(G/H_i). \tag{3.1}
$$

Here "$\cong$" means $\mathbb{Z}G$-isomorphism and G acts on $\bigoplus_{G/H_i} H_n(D^n, S^{n-1})$ by permuting the summands. Note that the isomorphism $H_n(D^n, S^{n-1}) \cong \mathbb{Z}$ is canonical because (D^n, S^{n-1}) has a preferred orientation coming from the standard orientation of $\mathbb{R}^n$. Thus $C_*(X)$ is of the form

$$
\cdots \longrightarrow \bigoplus_{i \in I_2} \mathbb{Z}(G/H_i) \longrightarrow \bigoplus_{i \in I_1} \mathbb{Z}(G/H_i) \longrightarrow \bigoplus_{i \in I_0} \mathbb{Z}(G/H_i) \longrightarrow 0. \tag{3.2}
$$

How much does the isomorphism $C_n(X) \cong \bigoplus_{i \in I_n} \mathbb{Z}(G/H_i)$ depend on the chosen pushout? Observe that the complete pushout diagram is determined by the lower map Q_n. The map Q_n, in turn, is G-equivariant and therefore determined by what it does on $\coprod_{i \in I_n} \{eH_i\} \times D^n$. So the choice of a pushout is in more concrete terms the choice of one n-cell in each G-orbit of n-cells together with its characteristic map. We want to refer to this data as a *cellular basis* of the G-CW complex X.

Proposition 3.6 *Let X be a G-CW complex. A cellular basis for X gives rise to $\mathbb{Z}G$-isomorphisms $\bigoplus_{i \in I_n} \mathbb{Z}(G/H_i) \cong C_*(X)$ where the H_i are the stabilizer groups of the chosen cells. We obtain the isomorphism of any other cellular basis by precomposing with*

$$\bigoplus_{i \in I_n} r_{\pm g_i H} : \bigoplus_{i \in I_n} \mathbb{Z}(G/g_i H_i g_i^{-1}) \xrightarrow{\sim} \bigoplus_{i \in I_n} \mathbb{Z}(G/H_i)$$

where $r_{\pm g_i H_i} : \mathbb{Z}(G/g_i H_i g_i^{-1}) \to \mathbb{Z}(G/H_i)$ is right multiplication with $\pm g_i H_i$. Here the $g_i H_i$ are the unique cosets moving the cells of the first basis to the cells of the second.

Proof Let $\phi_1, \phi_2 \colon D^n \to X_n$ be characteristic maps of two cellular bases which each pick out one n-cell in the same fixed G-orbit determined by $i \in I_n$. Then there is $g_i \in G$ with $g_i \phi_1(D^n) = \phi_2(D^n)$. Thus if $g_1 \phi_1(D^n) = g_2 \phi_2(D^n)$, then $g_1 \phi_1(D^n) = g_2 g_i \phi_1(D^n)$ which says $g_1 H_i = g_2 g_i H_i$ where H_i is the stabilizer group of the cell $\phi_1(D^n)$.

Let $[D^n, S^{n-1}] \in H_n(D^n, S^{n-1})$ be the canonical generator. Then the elements $H_n(g_i \phi_1)([D^n, S^{n-1}])$ and $H_n(\phi_2)([D^n, S^{n-1}])$ generate the same direct summand in the free $\mathbb{Z}$-module $H_n(X^n, X^{n-1})$. Thus $H_n(g_i \phi_1)([D^n, S^{n-1}]) = \pm H_n(\phi_2)([D^n, S^{n-1}])$.

So for every G-orbit of n-cells $i \in I_n$ we have an embedding $\mathbb{Z}(G/H_i) \hookrightarrow C_n(X)$ which is unique up to precomposition with an isomorphism of the form $\mathbb{Z}(G/g_i H_i g_i^{-1}) \xrightarrow{\sim} \mathbb{Z}(G/H_i)$ given by right multiplication with the coset $\pm g_i H_i$. This is indeed an isomorphism because right multiplication with $\pm H_i g_i^{-1}$ defines an inverse.   □

We will now see that under favorable circumstances the cellular chain complex $C_*(X)$ of a G-CW complex X can be completed to a chain complex of Hilbert $\mathcal{L}(G)$-modules by means of the functor $\ell^2 G \otimes_{\mathbb{Z}G} (\,\cdot\,)$. It should come as no surprise that circumstances are favorable if X is proper and finite type. This has the effect that the cellular differentials are $\mathbb{Z}G$-morphisms of the form

$$f \colon \bigoplus_{i \in I^1} \mathbb{Z}(G/H_i) \longrightarrow \bigoplus_{j \in I^2} \mathbb{Z}(G/H_j) \tag{3.3}$$

for finite families $\{H_i\}_{i \in I^1}$ and $\{H_j\}_{j \in I^2}$ of finite subgroups of G. We start with two plain algebraic-analytic propositions which explain the effect of the functor

$\ell^2 G \otimes_{\mathbb{Z}G} (\cdot)$ on objects and morphisms of this type. The statements should be plausible but the proofs require some thought. We consider $\ell^2 G$ as a $\mathbb{C}G$-$\mathbb{Z}G$-bimodule where $\mathbb{C}G$ acts from the left by λ and $\mathbb{Z}G$ acts from the right by $(\cdot)^* \circ \rho$.

Proposition 3.7 *Let $H \subset G$ be a finite subgroup. Then we have a canonical isomorphism of left $\mathbb{C}G$-modules $\ell^2 G \otimes_{\mathbb{Z}G} \mathbb{Z}(G/H) \cong \ell^2(G/H)$.*

Proof Let $n = |H|$. Requiring that the distinguished elements $e \otimes H \in \ell^2 G \otimes_{\mathbb{Z}G} \mathbb{Z}(G/H)$ and $H \in \ell^2(G/H)$ should correspond to one another determines two $\mathbb{C}G$-homomorphisms $\Phi: \ell^2 G \otimes_{\mathbb{Z}G} \mathbb{Z}(G/H) \to \ell^2(G/H)$ and $\Psi: \ell^2(G/H) \to \ell^2 G \otimes_{\mathbb{Z}G} \mathbb{Z}(G/H)$ uniquely as follows

$$\Phi: \left(\sum_{g \in G} c_g g \right) \otimes H \longmapsto \sum_{gH \in G/H} \left(\sum_{h \in H} c_{gh} \right) gH \tag{3.4}$$

$$\Psi: \sum_{gH \in G/H} c_{gH} gH \longmapsto \left(\sum_{gH \in G/H} \left(\frac{c_{gH}}{n} \sum_{h \in H} gh \right) \right) \otimes H. \tag{3.5}$$

Clearly $\Phi \circ \Psi = \text{id}$. But the reverse composition is tricky. First we get

$$\Psi \circ \Phi \left(\left(\sum_{g \in G} c_g g \right) \otimes H \right) = \left(\sum_{gH \in G/H} \frac{(\sum_{h \in H} c_{gh})}{n} \sum_{h \in H} gh \right) \otimes H.$$

In words, $\Psi \circ \Phi$ effects on the left factor $\sum c_g g$ of a simple tensor that every Fourier coefficient c_g is replaced by the mean of the Fourier coefficients throughout the coset gH. But since $gh \otimes H = g \otimes H$, we have

$$\left(\frac{(\sum_{h \in H} c_{gh})}{n} \sum_{h \in H} gh \right) \otimes H = \sum_{h \in H} c_{gh} gh \otimes H. \tag{3.6}$$

Thus for every finite subset $F \subseteq G/H$ we obtain

$$\left(\sum_{gH \in G/H} \frac{(\sum_{h \in H} c_{gh})}{n} \sum_{h \in H} gh \right) \otimes H = \tag{3.7}$$

$$= \left(\sum_{gH \in F \subseteq G/H} \sum_{h \in H} c_{gh} gh + \sum_{gH \in G/H \setminus F} \frac{(\sum_{h \in H} c_{gH})}{n} \sum_{h \in H} gh \right) \otimes H. \tag{3.8}$$

It is now tempting to say that this equals $\left(\sum_{g \in G} c_g g \right) \otimes H$ by passing to the limit for larger and larger F. While in the end this will be true, the assertion does not make sense at this point because we have not defined any topology on

$\ell^2 G \otimes_{\mathbb{Z}G} \mathbb{Z}(G/H)$ yet. To do so we observe that the canonical $\mathbb{Z}G$-homomorphism $q \colon \ell^2 G \to \ell^2 G \otimes_{\mathbb{Z}G} \mathbb{Z}(G/H)$ given by $x \mapsto x \otimes H$ is surjective. So we assign the finest topology to $\ell^2 G \otimes_{\mathbb{Z}G} \mathbb{Z}(G/H)$ for which q is still continuous. We want to show that this gives a T_1-space (points are closed). Since q is a quotient map by construction, we have to show that preimages of points under q are closed in $\ell^2 G$. Preimages of points under q are precisely the affine subspaces over $\ker q$. By (3.6) the subspace $\ker q$ consists of all Fourier series whose Fourier coefficients sum up to zero over left H-cosets. But this space is just the orthogonal complement of the image of the canonical inclusion $\ell^2(G/H) \hookrightarrow \ell^2 G$ constructed in Example 2.41 and thus is closed.

The point of these remarks is that in T_1-spaces constant nets have unique limits. Thus the element in (3.7) is the unique limit of the constant net given in (3.8), directed over finite subsets $F \subseteq G/H$. By continuity of q this limit equals $\left(\sum_{g \in G} c_g g \right) \otimes H$. Whence $\Psi \circ \Phi = \mathrm{id}$.   □

The families $\{H_i\}_{i \in I^1}$ and $\{H_j\}_{j \in I^2}$ from (3.3) determine isomorphisms $\{\Phi_i\}_{i \in I^1}$ and $\{\Psi_j\}_{j \in I^2}$ as in (3.4) and (3.5). We want to refer to the $\mathbb{C}G$-homomorphism

$$f^{(2)} \colon \bigoplus_{i \in I^1} \ell^2(G/H_i) \longrightarrow \bigoplus_{j \in I^2} \ell^2(G/H_j). \qquad (3.9)$$

given by $f^{(2)} = \left(\bigoplus_{i \in I^1} \Phi_i \right) \circ (\mathrm{id} \otimes f) \circ \left(\bigoplus_{j \in I^2} \Psi_j \right)$ as the ℓ^2-extension of f.

Proposition 3.8 *For every $\mathbb{Z}G$-homomorphism f as in (3.3) the ℓ^2-extension $f^{(2)}$ in (3.9) is a bounded operator of Hilbert spaces. It is given by right multiplication with the matrix $M_{ij} = (f(H_i))_j \in \mathbb{Z}(G/H_j)^{H_i}$ according to the well-defined rule $g H_i M_{ij} = g M_{ij}$.*

Proof Let $M(f)_i = f(H_i) \in \bigoplus_{j \in I^2} \mathbb{Z}(G/H_j)$ be the image of the H_i-invariant element $H_i \in \mathbb{Z}(G/H_i)$ under f. By G-equivariance of f, the element $M(f)_i$ is likewise H_i-invariant and the same goes for all the components $M(f)_{ij} \in \mathbb{Z}(G/H_j)$ of $M(f)_i \in \bigoplus_{j \in I^2} \mathbb{Z}(G/H_j)$ because G acts diagonally. Let $\overline{M(f)_{ij}}$ be any lift of $M(f)_{ij}$ under $\mathbb{Z}G \to \mathbb{Z}(G/H_j)$. Then for $x \in \ell^2 G$ we obtain

$$(\mathrm{id} \otimes f)(x \otimes H_i) = x \otimes M(f)_i = \sum_{j \in I^2} x \otimes \overline{M(f)_{ij}} H_j = \sum_{j \in I^2} x \overline{M(f)_{ij}} \otimes H_j.$$

It follows that the homomorphism $f^{(2)}$ is given by right multiplication with the matrix $M(f)_{ij}$ applying the well-defined rule $g H_i M(f)_{ij} = g M(f)_{ij}$.

It remains to see that this gives a bounded operator. To this end let $H \leq G$ be a finite subgroup and let $\pi \colon \ell^2 G \to \ell^2(G/H)$ be the canonical operator given by $g \mapsto gH$. Composing $\frac{\pi}{\sqrt{|H|}}$ with the canonical isometric embedding $\ell^2(G/H) \hookrightarrow \ell^2 G$ from Example 2.41 we obtain the orthogonal projection onto the closed subspace of $\ell^2 G$ consisting of elements with constant Fourier coefficients throughout H-cosets. Thus $\|\pi\| = \sqrt{|H|}$. Now let S_i be a system of representatives for the cosets G/H_i, let $x_i = \sum_{g_1 \in S_i} c_{g_1 H_i}\, g_1 H_i$ be some element in $\ell^2(G/H_i)$ and for fixed i and j write the matrix entry $M_{ij} \in \mathbb{Z}(G/H_j)^{H_i}$ as $M_{ij} = \sum_{g_2 H_j} d_{g_2 H_j}\, g_2 H_j$ where almost all coefficients $d_{g_2 H_j} \in \mathbb{Z}$ vanish. Then considering the above we obtain

$$\|x_i \cdot M_{ij}\| = \left\|\left(\sum_{g_1 \in S_i} c_{g_1 H_i}\, g_1 H_i\right) M_{ij}\right\| = \left\|\sum_{g_1 \in S_i} c_{g_1 H_i} g_1 M_{ij}\right\|$$

$$= \left\|\sum_{g_2 H_j} d_{g_2 H_j} \sum_{g_1 \in S_i} c_{g_1 H_i}\, g_1 g_2 H_j\right\| \le \sum_{g_2 H_j} |d_{g_2 H_j}| \left\|\sum_{g_1 \in S_i} c_{g_1 H_i}\, g_1 g_2 H_j\right\|$$

$$\le \|M_{ij}\|_1 \sqrt{|H_j|}\|x_i\|$$

with $\|M_{ij}\|_1 := \sum_{g_2 H_j} |d_{g_2 H_j}|$. Therefore we can estimate the norm of an element $(x_i)_{i \in I^1} \in \bigoplus_{i \in I^1} \ell^2(G/H_i)$ multiplied from the right by M as

$$\|(x_i)_{i \in I^1} \cdot M\|^2 = \left\|\left(\sum_{i \in I^1} x_i \cdot M_{ij}\right)_{j \in I^2}\right\|^2 = \sum_{j \in I^2} \left\|\sum_{i \in I^1} x_i \cdot M_{ij}\right\|^2$$

$$\le \sum_{j \in I^2} \left(\sum_{i \in I^1} \|x_i \cdot M_{ij}\|\right)^2 \le \sum_{j \in I^2} \left(\sum_{i \in I^1} \|M_{ij}\|_1 \sqrt{|H_j|}\|x_i\|\right)^2$$

$$\le \left(|I^2| \cdot |I^1|^2 \max_{j \in I^2}\{|H_j|\} \cdot \|M\|_1^2\right) \|(x_i)_{i \in I^1}\|^2$$

$$= \text{const} \cdot \|(x_i)_{i \in I^1}\|^2$$

where $\|M\|_1 := \max_{ij} \|M_{ij}\|_1$. Whence $f^{(2)}$ is a bounded operator. $\square$

In the context of this proposition it is convenient to observe that the $\mathbb{Z}$-submodule of $\mathbb{Z}(G/H_j)$ consisting of H_i-invariant elements can be described as $\mathbb{Z}(G/H_j)^{H_i} = \mathbb{Z}(h_i\, G/H_j)$ where $h_i = \sum_{h \in H_i} h \in \mathbb{Z}G$ is the canonical H_i-invariant element. Indeed, for $x = \sum_{g H_j} c_{g H_j} g H_j \in \mathbb{Z}(G/H_j)$ and $h \in H_i$ we obtain that $hx = x$ is equivalent to $c_{h^{-1} g H_j} = c_{g H_j}$ for all $g H_j \in G/H_j$. So the matrix M from above has entries M_{ij} in $\mathbb{Z}(h_i\, G/H_j)$ which once more explains the rule $g H_i M_{ij} = g M_{ij}$. Note that the $\mathbb{Z}$-submodule $\mathbb{Z}(h_i\, G/H_j)$ of $\mathbb{Z}(G/H_j)$ is *dual* to the $\mathbb{Z}$-submodule $\mathbb{Z}(h_j G/H_i)$ of $\mathbb{Z}(G/H_i)$ under the well-defined $*$-operation $(h_i g H_j)^* = h_j g^{-1} H_i$ and $(h_j g H_i)^* = h_i g^{-1} H_j$.

Proposition 3.9 *The Hilbert space adjoint $f^{(2)*}$ of $f^{(2)}$ is given by right multiplication with the matrix $(M^*)_{ji} := (M_{ij})^*$.*

Proof This is a pure calculational matter along the canonical orthonormal bases $\coprod_{i\in I_1} G/H_i$ and $\coprod_{j\in I_2} G/H_j$. We have

$$\langle (g_1 H_i) h_i g H_j, g_2 H_j \rangle = \langle g_1 h_i g H_j, g_2 H_j \rangle = \sum_{h^i \in H_i} \langle g_1 h^i g H_j, g_1 H_j \rangle$$

$$= \sum_{h^i \in H_i} \sum_{h^j \in H_j} \langle g_1 h^i g, g_2 h^j \rangle = \sum_{h^j \in H_j} \sum_{h^i \in H_i} \langle g_1 h^i, g_2 h^j g^{-1} \rangle$$

$$= \sum_{h^j \in H_j} \langle g_1 H_i, g_2 h^j g^{-1} H_i \rangle = \langle g_1 H_i, g_2 h_j g^{-1} H_i \rangle$$

$$= \langle g_1 H_i, (g_2 H_j) h_j g^{-1} H_i \rangle. \qquad \square$$

Now we feel well prepared to study the ℓ^2-completion of a cellular chain complex.

Definition 3.10 The ℓ^2-*chain complex* of a G-CW complex X is the $\mathbb{C}G$-chain complex given by $C_*^{(2)}(X) = \ell^2 G \otimes_{\mathbb{Z}G} C_*(X)$.

Thus the ℓ^2-chain complex construction is the composition of the cellular chain complex C_*, which is functorial by Proposition 3.5, and the tensor functor $\ell^2 G \otimes_{\mathbb{Z}G} (\cdot)$. In particular, the differentials are given by $d_*^{(2)} = \mathrm{id} \otimes d_*$ where d_* is the cellular differential. Since $(\mathrm{id} \otimes d) \circ (\mathrm{id} \otimes d) = \mathrm{id} \otimes d^2 = 0$, we obtain a functor $(C_*^{(2)}, d_*^{(2)})$ from G-CW complexes to $\mathbb{C}G$-chain complexes. But on the (full) subcategory of proper, finite type G-CW complexes, something better is true.

Theorem 3.11 *The ℓ^2-chain complex defines a functor $(C_*^{(2)}, d_*^{(2)})$ from proper, finite type G-CW complexes to chain complexes of Hilbert $\mathcal{L}(G)$-modules.*

Proof Let X be a proper, finite type G-CW complex and pick a cellular basis for X. This determines finite families of finite stabilizer subgroups $\{H_i\}_{i\in I_n}$. Equation (3.1) and Proposition 3.7 therefore combine to give an isomorphism $C_n^{(2)}(X) \cong \bigoplus_{i\in I_n} \ell^2(G/H_i)$. We pull back the inner product of $\bigoplus_{i\in I_n} \ell^2(G/H_i)$ along this isomorphism to turn $C_n^{(2)}(X)$ into a Hilbert space with isometric, linear left G-action. The isomorphisms of Proposition 3.6 become G-equivariant unitaries under the $\ell^2(G) \otimes_{\mathbb{Z}G}$-functor so that the Hilbert space structure on $C_n^{(2)}(X)$ is independent of the cellular basis. Thus Example 2.41 gives an embedding $\bigoplus_{i\in I_n} \ell^2(G/H_i) \hookrightarrow (\ell^2 G)^{k_n}$, where $k_n = |I_n|$. This verifies that $C_n^{(2)}(X)$ has a canonical structure of a Hilbert $\mathcal{L}(G)$-module.

It remains to establish that the differentials $d_n^{(2)}$ and the $\mathbb{C}G$-morphism $C_n^{(2)}(f)$ of a G-equivariant, cellular map $f : X \to Y$ of proper, finite type G-CW complexes X and Y are bounded operators. But this is what Proposition 3.8 asserts after applying the isomorphism (3.1) for X and Y coming from any cellular bases. $\qquad \square$

The situation is particularly transparent when the proper G-CW complex X is actually free. Note that a proper G-CW complex is automatically free if the group G is torsion-free. In the case of a free, finite type G-CW complex X the ℓ^2-chain complex $C_*^{(2)}(X)$ consists of free Hilbert $\mathcal{L}(G)$-modules so that it is of the form

$$\cdots \longrightarrow (\ell^2 G)^{k_2} \longrightarrow (\ell^2 G)^{k_1} \longrightarrow (\ell^2 G)^{k_0} \longrightarrow 0.$$

Proposition 3.8 says in this case that the differentials are given by right multiplication with matrices over the integral group ring $\mathbb{Z}G$. Proposition 3.9 says that the adjoints of the differentials are given by right multiplication with the transposed matrices whose entries are moreover involuted by the canonical ring involution of $\mathbb{Z}G$ given by $g \mapsto g^{-1}$.

Exercises

3.2.1 Let G be a group and let $H \leq G$ be a finite subgroup. Show that the $\mathbb{Z}$-submodule $(\mathbb{Z}G)^H$ of H-invariant elements in $\mathbb{Z}G$ is a $\mathbb{Z}G$-submodule if and only if H is a normal subgroup.

3.2.2 Let Y be the D_∞-CW complex from Exercise 3.1.3. Show that $d_1 \colon C_1^{(2)}(Y) \to C_0^{(2)}(Y)$ is a weak isomorphism by proving that it is injective and that $\dim_{\mathcal{R}(G)} C_1^{(2)}(Y) = \dim_{\mathcal{R}(G)} C_0^{(2)}(Y)$.

3.3 ℓ^2-Betti Numbers and How to Compute Them

Our journey arrives at a milestone. We are in the position to give the definitions we have been longing for: ℓ^2-homology and ℓ^2-Betti numbers. Afterwards we look at some concrete and very basic examples for which we can make computations directly from the ℓ^2-chain complex in order to acquire some familiarity with the situation. Only then will we move on to study properties of ℓ^2-Betti numbers systematically.

Definition 3.12 Let X be a proper, finite type G-CW complex with ℓ^2-chain complex $(C_*^{(2)}(X), d_*^{(2)})$. The *n-th (reduced) ℓ^2-homology* of X is the Hilbert $\mathcal{L}(G)$-module $H_n^{(2)}(X) = \ker d_n^{(2)} / \overline{\operatorname{im} d_{n+1}^{(2)}}$.

Let us ponder for a moment why this definition is meaningful: The chain module $C_n^{(2)}(X)$ is a Hilbert module by Theorem 3.11. The kernel $\ker d_n^{(2)}$ is a closed G-invariant subspace because $d_n^{(2)}$ is continuous and G-equivariant. So $\ker d_n^{(2)}$ is a Hilbert submodule by the discussion below Theorem 2.35. The image $\operatorname{im} d_{n+1}^{(2)}$ is a G-invariant subspace of $\ker d_n^{(2)}$ but it might not be closed: we will see in a minute that the operator of Example 2.42 occurs as $d_1^{(2)}$ in $C^{(2)}(\widetilde{S^1})$! We thus take the closure before going over to the quotient. In this manner $H_n^{(2)}(X)$ is a well-defined *Hilbert subquotient* of $C_n^{(2)}(X)$. We can also consider the ordinary *unreduced* ℓ^2-homology $H_{n,\mathrm{unr}}^{(2)}(X) = \ker d_n^{(2)} / \operatorname{im} d_{n+1}^{(2)}$ as a quotient of $\mathbb{C}G$-modules but again, this object generally comes with no natural Hilbert module structure.

Definition 3.13 Let X be a proper, finite type G-CW complex. The *n-th ℓ^2-Betti number* of X is given by $b_n^{(2)}(X) = \dim_{\mathcal{R}(G)} H_n^{(2)}(X)$.

So by definition ℓ^2-Betti numbers are nonnegative real numbers. Be aware that $b_n^{(2)}(X)$ depends crucially on the action of G on X. To capture this dependence in the notation, we will occasionally write $b_n^{(2)}(G \curvearrowright X)$ instead of $b_n^{(2)}(X)$. In the special case of a Galois covering $\overline{X}$ of a finite CW complex X, it is understood that $b_n^{(2)}(\overline{X})$ means $b_n^{(2)}(G \curvearrowright \overline{X})$ where $G \curvearrowright \overline{X}$ is the deck transformation action. In particular, for any connected finite CW complex X, the notation $b_n^{(2)}(\widetilde{X})$ means $b_n^{(2)}(\pi_1 X \curvearrowright \widetilde{X})$.

Example 3.14 Let G be a finite group. Then every G-CW complex X is proper. Moreover X is of finite type if and only if all skeleta are compact. In this case, the ℓ^2-chain complex

$$C_*^{(2)}(X) = \ell^2 G \otimes_{\mathbb{Z}G} C_*(X) \cong \mathbb{C}G \otimes_{\mathbb{Z}G} C_*(X) \cong (\mathbb{C} \otimes_{\mathbb{Z}} \mathbb{Z}G) \otimes_{\mathbb{Z}G} C_*(X) \cong$$

$$\cong \mathbb{C} \otimes_{\mathbb{Z}} (\mathbb{Z}G \otimes_{\mathbb{Z}G} C_*(X)) \cong \mathbb{C} \otimes_{\mathbb{Z}} C_*(X) \cong C_*(X; \mathbb{C})$$

is just the cellular chain complex with complex coefficients. Note that associativity of tensor products also holds with different rings involved [24, Chapter II, Section 3.8]. So ℓ^2-homology for finite G equals ordinary homology and we obtain from Example 2.39 that $b_n^{(2)}(G \curvearrowright X) = \frac{b_n(X)}{|G|} \in \frac{1}{|G|}\mathbb{Z}_{\geq 0}$ where $b_n(X) = \dim_{\mathbb{C}} H_n(X; \mathbb{C})$ is the classical n-th Betti number. For G the trivial group, in particular, a G-CW complex is the same as an ordinary CW complex and ℓ^2-Betti numbers reduce to ordinary Betti numbers.

The example reveals that ℓ^2-Betti numbers report nothing new if G is finite. But this is not a bug; it's a feature! ℓ^2-invariants are designed as an *extension* of the classical theory to infinite groups.

Example 3.15 Let $X = \widetilde{S^1}$ be the $\mathbb{Z}$-CW complex given by the universal covering of the circle S^1 with the standard CW structure consisting of one 0-cell and one 1-cell. So X is an infinite line built from one free $\mathbb{Z}$-equivariant 0-cell and one free $\mathbb{Z}$-equivariant 1-cell. A choice of a cellular basis is indicated by the thickened 0- and 1-cell in the following image.

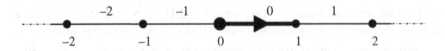

These determine the labeling of all cells by elements of $\mathbb{Z}$. Part of the cellular basis is the characteristic map $D^1 \to X$ of the chosen 1-cell which equips this cell with an orientation. Say this orientation is the one indicated by the arrow. Then the first cellular differential d_1 maps the chosen 1-cell to the 0-cell labeled "1" minus the 0-cell labeled "0". Thus (3.1), Proposition 3.8 and Fourier transform as in Example 2.14 realize the ℓ^2-differential $d_1^{(2)}$ as the operator $L^2[-\pi, \pi] \to L^2[-\pi, \pi]$ given by multiplication with the function $(z - 1)$ where $z = e^{ix}$ for $x \in [-\pi, \pi]$. As we saw in Example 2.42 this is a weak isomorphism, thus $H_0^{(2)}(X) = C_0^{(2)}(X)/\overline{\operatorname{im} d_1^{(2)}} = 0$ and $H_1^{(2)}(X) = \ker d_1^{(2)} = 0$. Since X has no cells in dimensions larger than one, it follows that X is ℓ^2-*acyclic*: we have $b_n^{(2)}(X) = 0$ for all $n \geq 0$. Note however that $H_{0,\mathrm{unr}}^{(2)}(X) \neq 0$. We remark that this phenomenon can be captured by so-called *Novikov–Shubin invariants* [117, Chapter 2].

Example 3.16 Let Y be the CW complex in the following picture

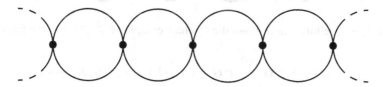

which can formally be defined by the pushout diagram

$$
\begin{array}{ccc}
X_0 & \longrightarrow & X \\
\downarrow & & \downarrow \\
X & \longrightarrow & Y
\end{array}
$$

where X is the CW complex from above and X_0 is the 0-skeleton. The group $G = \mathbb{Z} \times \mathbb{Z}/2\mathbb{Z}$ acts cellularly on Y in the only natural way: the $\mathbb{Z}$-factor by translation and the $\mathbb{Z}/2\mathbb{Z}$-factor by swapping upper and lower arcs. As a G-CW complex Y is finite and proper but not free. It has one equivariant 0-cell with stabilizer $\{0\} \times \mathbb{Z}/2\mathbb{Z}$ and one free equivariant 1-cell. A cellular basis identifies the ℓ^2-chain complex with

$$\cdots \longrightarrow 0 \longrightarrow \ell^2(\mathbb{Z} \times \mathbb{Z}/2\mathbb{Z}) \xrightarrow{d_1^{(2)}} \ell^2(\mathbb{Z} \times \mathbb{Z}/2\mathbb{Z} / \{0\} \times \mathbb{Z}/2\mathbb{Z}) \longrightarrow 0.$$

Since X is included in Y as a subcomplex and $X_0 = Y_0$, the differential $d_1^{(2)}$ must again have dense image so that $b_0^{(2)}(Y) = 0$. It follows that $b_1^{(2)}(Y) = \dim_{\mathcal{R}(\mathbb{Z} \times \mathbb{Z}/2\mathbb{Z})} \ker d_1^{(2)}$ is given by

$$\dim_{\mathcal{R}(\mathbb{Z} \times \mathbb{Z}/2\mathbb{Z})} \ell^2(\mathbb{Z} \times \mathbb{Z}/2\mathbb{Z}) - \dim_{\mathcal{R}(\mathbb{Z} \times \mathbb{Z}/2\mathbb{Z})} \ell^2(\mathbb{Z} \times \mathbb{Z}/2\mathbb{Z} / \{0\} \times \mathbb{Z}/2\mathbb{Z})$$

$$= 1 - 0.5 = 0.5$$

where we applied additivity of von Neumann dimension (Theorem 2.44 (v)) and the calculation in Example 2.41. Again there are no higher-dimensional cells so we have $b_n^{(2)}(Y) = 0$ for all $n \geq 2$.

Example 3.17 Now we want to fill in 2-cells into the circles of Y. But if we want that $G = \mathbb{Z} \times \mathbb{Z}/2\mathbb{Z}$ still acts cellularly, this requires us to also include some new 1-cells to account for the fixed point sets of the subgroup $F = \{0\} \times \mathbb{Z}/2\mathbb{Z}$ which is still supposed to flip upper and lower half of the complex. Thus the new G-CW complex Z contains Y as a G-subcomplex and additionally has one equivariant 1-cell with stabilizer F and one free equivariant 2-cell.

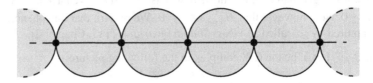

Accordingly, a cellular basis shows the ℓ^2-chain complex $C_*^{(2)}(Z)$ is of the form

$$\cdots \longrightarrow 0 \longrightarrow \ell^2 G \xrightarrow{d_2^{(2)}} \ell^2 G \oplus \ell^2(G/F) \xrightarrow{d_1^{(2)}} \ell^2(G/F) \longrightarrow 0.$$

Note however that the differentials are not the inclusion and projection of the direct summands. The first differential $d_1^{(2)}$ has dense image as it already does when restricted to the submodule coming from the subcomplex Y. The second differential $d_2^{(2)}$ is injective because the first component of $d_2^{(2)}$ is $\mathrm{id}_{\ell^2 G}$. Since the von Neumann dimensions of the outer terms add up to 1.5 which is the von Neumann dimension of the middle term, it follows that the ℓ^2-chain complex is weakly exact and thus Z is ℓ^2-acyclic.

Of course the rules of the game are to avoid chain complex considerations whenever possible. Instead, the following properties provide more systematic methods to compute ℓ^2-Betti numbers of proper, finite type G-CW complexes.

Theorem 3.18 (Computation of ℓ^2-Betti Numbers)

(i) Homotopy invariance. *Let $f : X \to Y$ be a G-homotopy equivalence of proper, finite type G-CW complexes X and Y. Then $b_n^{(2)}(X) = b_n^{(2)}(Y)$ for all $n \geq 0$.*

(ii) Zeroth ℓ^2-Betti number. *Let X be a connected, nonempty, proper, finite type G-CW complex. Then $b_0^{(2)}(X) = \frac{1}{|G|}$ with $\frac{1}{\infty} = 0$.*

(iii) Künneth formula. *Let X_1 and X_2 be a proper, finite type G_1- and G_2-CW complexes, respectively. Then $X_1 \times X_2$ is a proper, finite type $G_1 \times G_2$-CW complex and for all $n \geq 0$ we have*

$$b_n^{(2)}(X_1 \times X_2) = \sum_{p+q=n} b_p^{(2)}(X_1) \, b_q^{(2)}(X_2).$$

(iv) Restriction. *Let X be a proper, finite type G-CW complex and let $G_0 \le G$ be a finite index subgroup. Then $\text{res}_{G_0}^G X$ is a proper, finite type G_0-CW complex and $b_n^{(2)}(\text{res}_{G_0}^G X) = [G : G_0]b_n^{(2)}(X)$ for all $n \ge 0$.*

Proof We start with (i). As a consequence of Theorem 3.11 we see that $H_n^{(2)}$ is a functor from proper, finite type G-CW complexes to Hilbert $\mathcal{L}(G)$-modules and we have to show that this functor factorizes over the homotopy category of G-CW complexes. In other words, if the cellular G-maps $f_0, f_1 \colon X \to Y$ are homotopic by a homotopy $h \colon X \times I \to Y$ through cellular G-maps h_t with $h_0 = f_0$ and $h_1 = f_1$, then $H_n^{(2)}(f_0) = H_n^{(2)}(f_1)$. For a moment let's take this fact for granted. By cellular approximation we can assume that f, its G-homotopy inverse and the homotopies to the identity are cellular. It then follows from the above that $H_n^{(2)}(f)$ is an isomorphism of Hilbert modules. Note however that Hilbert module morphisms are not required to be isometric so that $H_n^{(2)}(f)$ must not be unitary. Thus it is not quite immediate that $H_n^{(2)}(X)$ and $H_n^{(2)}(Y)$ have equal von Neumann dimension. Nevertheless, (weakly) isomorphic Hilbert modules V and W do have equal von Neumann dimension because we can argue that

$$0 \to 0 \to V \xrightarrow{\sim} W \to 0$$

is (weakly) exact so that $\dim_{\mathcal{R}(G)} V = \dim_{\mathcal{R}(G)} W$ follows from additivity, Theorem 2.44 (v). Let us now prove $H_n^{(2)}(f_0) = H_n^{(2)}(f_1)$. The homotopy h is cellular so that $h((X \times I)_n) \subseteq Y_n$ where $(X \times I)_n = X_n \times \partial I \cup X_{n-1} \times I$. Since it is also G-equivariant, the induced collection of maps $\gamma_n \colon C_n(X) \to C_{n+1}(Y)$ given by

$$H_n(X_n, X_{n-1}) \xrightarrow{\sim} H_{n+1}(X_n \times I, X_n \times \partial I \cup X_{n-1} \times I) \xrightarrow{H_{n+1}(h)} H_{n+1}(Y_{n+1}, Y_n)$$

consists of $\mathbb{Z}G$-homomorphisms, where the first map is suspension. One checks that γ_* is actually a chain homotopy from $C_*(f_1)$ to $C_*(f_2)$: we have $C_*(f_1) - C_*(f_2) = d_{*+1}\gamma_* + \gamma_{*-1}d_*$ as proven for example in [167, Proposition 12.1.6, p. 303]. Applying $\ell^2 G \otimes_{\mathbb{Z}G} (\cdot)$ we obtain $C_*^{(2)}(f_1) - C_*^{(2)}(f_2) = d_{*+1}^{(2)}\gamma_*^{(2)} + \gamma_{*-1}^{(2)}d_*^{(2)}$ which maps ℓ^2-cycles to ℓ^2-boundaries so $H_n^{(2)}(f_1) = H_n^{(2)}(f_2)$ as desired.

Part (ii) in case G is a finite group follows immediately from Example 3.14 because X is connected. The case of an infinite group is most naturally proven with the concept of classifying spaces available so that we postpone the proof to Sect. 4.5.4 of Chap. 4 on p. 83.

To see (iii) we observe that we have an isomorphism

$$C_*(X_1; \mathbb{C}) \otimes C_*(X_2; \mathbb{C}) \xrightarrow{\sim} C_*(X_1 \times X_2; \mathbb{C})$$

of $\mathbb{C}(G_1 \times G_2)$-chain complexes which maps a basis vector $c_p \otimes c_q$ from the n-chains $\bigoplus_{p+q=n} C_p(X_1; \mathbb{C}) \otimes_{\mathbb{C}} C_q(X_2; \mathbb{C})$ to the product cell $c_p \times c_q$ in $C_n(X_1 \times$

X_2; $\mathbb{C}$). Note that the corresponding situation for singular chain complexes is way less convenient: only a chain homotopy equivalence is available whose inverse must be constructed by the abstract method of acyclic models. We have inclusions

$$
\begin{array}{ccc}
C_*^{(2)}(X_1) \otimes C_*^{(2)}(X_2) - \xrightarrow{\ \exists!\ } & C_*^{(2)}(X_1 \times X_2) \\
\text{dense} \Big\uparrow & \text{dense} \Big\uparrow \\
C_*(X_1; \mathbb{C}) \otimes C_*(X_2; \mathbb{C}) \xrightarrow{\ \sim\ } & C_*(X_1 \times X_2; \mathbb{C})
\end{array}
$$

where the top left entry is a tensor product of chain complexes of Hilbert modules. So our isomorphism embeds a dense subspace of $C_*^{(2)}(X_1) \otimes C_*^{(2)}(X_2)$ isometrically into a dense subspace of $C_*^{(2)}(X_1 \times X_2)$. It follows from Exercise 2.2.2 that this embedding extends uniquely to a unitary isomorphism of Hilbert modules as indicated in the diagram. Thus what we still need is a Künneth type theorem saying that for chain complexes of Hilbert modules (C_*, c_*) and (D_*, d_*) we have an isomorphism

$$
\bigoplus_{p+q=n} H_n^{(2)}(C_*) \otimes H_n^{(2)}(D_*) \xrightarrow{\ \sim\ } H_n^{(2)}(C_* \otimes D_*).
$$

We content ourselves with pointing out the key reason why this works. The homology $H_n^{(2)}(C_*)$ can be identified with the orthogonal complement of $\mathrm{im}(d_{n+1})$ in $\ker d_n$. Since C_n is a Hilbert module, it embeds into some $(\ell^2 G)^N$ so that $H_n^{(2)}(C_*)$ is a direct summand in a free Hilbert module and in this sense is "projective". Thus there are no "Tor" phenomena. For the technical details, consult [117, Lemma 1.22, p. 28]. Additivity and multiplicativity of von Neumann dimension (Theorem 2.44 (v) and (vi)) finish the proof of (iii).

To show part (iv) recall that X and $\mathrm{res}_{G_0}^G X$ are equal as (nonequivariant) CW complexes, so that $C_*(X)$ and $C_*(\mathrm{res}_{G_0}^G X)$ are equal as $\mathbb{Z}$-modules. Since the G-action on X permutes the cells and thus the canonical $\mathbb{Z}$-basis of $C_*(X)$, we obtain a natural isomorphism $C_*(\mathrm{res}_{G_0}^G X) \cong \mathrm{res}_{G_0}^G C_*(X)$ of chain complex of $\mathbb{Z}G_0$-modules. Applying $\ell^2 G_0 \otimes_{\mathbb{Z}G_0} (\cdot)$, this gives a natural isomorphism

$$
C_*^{(2)}(\mathrm{res}_{G_0}^G X) \cong \ell^2 G_0 \otimes_{\mathbb{Z}G_0} \mathrm{res}_{G_0}^G C_*(X) \cong \mathrm{res}_{G_0}^G C_*^{(2)}(X)
$$

of chain complexes of Hilbert $\mathcal{L}(G_0)$-modules. The reduced homology of the latter is isomorphic to $\mathrm{res}_{G_0}^G H_n^{(2)}(X)$. Thus $H_n^{(2)}(\mathrm{res}_{G_0}^G X) \cong \mathrm{res}_{G_0}^G H_n^{(2)}(X)$ and part (iv) follows from the restriction property of von Neumann dimension, Theorem 2.44 (vii). $\square$

If a proper G-CW complex X is not only finite type but honestly finite, we can consider the alternating sum of ℓ^2-Betti numbers $\chi^{(2)}(X) = \sum_{n \geq 0} (-1)^n b_n^{(2)}(X)$. By the above theorem, $\chi^{(2)}(X)$ is a homotopy invariant which, it turns out, can be read off directly from the G-CW structure of X.

Theorem 3.19 (ℓ^2-Euler-Poincaré Formula) *Let X be a proper, finite G-CW complex. For $n \geq 0$ let $\{H_i\}_{i \in I_n}$ be the family of stabilizer subgroups of the G-equivariant n-cells of X (unique up to conjugation). Then we have*

$$\chi^{(2)}(X) = \sum_{n \geq 0} (-1)^n \sum_{i \in I_n} \frac{1}{|H_i|}.$$

In particular, if X is free, we have $\chi^{(2)}(X) = \chi(G \backslash X)$.

Proof Since every equivariant n-cell in X gives one $\ell^2(G/H_i)$-summand in $C_n^{(2)}(X)$, we get $\dim_{\mathcal{R}(G)} C_n^{(2)}(X) = \sum_{i \in I_n} \frac{1}{|H_i|}$ from Example 2.41. Applying additivity of von Neumann dimension (Theorem 2.44 (v)) to the two short weakly exact sequences of Hilbert modules

$$0 \longrightarrow \ker d_n^{(2)} \longrightarrow C_n^{(2)}(X) \xrightarrow{d_n^{(2)}} \overline{\operatorname{im} d_n^{(2)}} \longrightarrow 0,$$

$$0 \longrightarrow \overline{\operatorname{im} d_{n+1}^{(2)}} \longrightarrow \ker d_n^{(2)} \longrightarrow H_n^{(2)}(X) \longrightarrow 0,$$

we obtain

$$\sum_{n \geq 0} (-1)^n \sum_{i \in I_n} \frac{1}{|H_i|} = \sum_{n \geq 0} (-1)^n \dim_{\mathcal{R}(G)} C_n^{(2)}(X)$$

$$= \sum_{n \geq 0} (-1)^n \left(\dim_{\mathcal{R}(G)} \ker d_n^{(2)} + \dim_{\mathcal{R}(G)} \overline{\operatorname{im} d_n^{(2)}} \right)$$

$$= \sum_{n \geq 0} (-1)^n \left(\dim_{\mathcal{R}(G)} \overline{\operatorname{im} d_{n+1}^{(2)}} + \dim_{\mathcal{R}(G)} H_n^{(2)}(X) \right.$$

$$\left. + \dim_{\mathcal{R}(G)} \overline{\operatorname{im} d_n^{(2)}} \right).$$

Since $d_0^{(2)} = 0$, the outer two summands telescope out and the term reduces to $\sum_{n \geq 0} (-1)^n b_n^{(2)}(X) = \chi^{(2)}(X)$. If X is free, we always have $|H_i| = 1$ so that the formula gives $\chi^{(2)}(X) = \sum_{n \geq 0} (-1)^n |I_n| = \chi(G \backslash X)$.   □

We have arrived at the first motivating result from the introduction.

Corollary 3.20 *The Singer conjecture (Conjecture 1.5) implies the Hopf conjecture (Conjecture 1.4).*

Proof The Singer conjecture asserts that the universal covering $\widetilde{M}$ of a closed, aspherical $2n$-dimensional manifold M can only have a nonzero ℓ^2-Betti number in degree n. If this is true, then

$$(-1)^n \chi(M) = (-1)^n \chi^{(2)}(\widetilde{M}) = (-1)^n (-1)^n b_n^{(2)}(\widetilde{M}) = b_n^{(2)}(\widetilde{M}) \geq 0.$$

□

With the properties established so far we can easily recover the calculated ℓ^2-Betti numbers of the G-CW complexes X, Y and Z from Examples 3.15–3.17. Each of them has vanishing zeroth ℓ^2-Betti number by Theorem 3.18 (ii). The circle has vanishing Euler characteristic so that $b_1^{(2)}(X) = 0$ follows from the ℓ^2-Euler-Poincaré formula, Theorem 3.19. The ℓ^2-Euler-Poincaré formula also gives $b_1^{(2)}(Y) = 0.5$. Alternatively, we can restrict the group action on Y to $\mathbb{Z} \times \{0\}$ which acts freely. The quotient space is homeomorphic to $S^1 \vee S^1$ which has Euler characteristic -1. Thus $b_1^{(2)}(Y) = 0.5$ follows from the restriction property, Theorem 3.18 (iv). Finally, Z and X are G-homotopy equivalent so that homotopy invariance, Theorem 3.18 (i), and the above give that Z is likewise ℓ^2-acyclic.

Exercises

3.3.1 Let F_k be the free group on $k \geq 2$ letters and let X be the free, finite F_k-CW complex given by the universal covering of a wedge of k circles with the standard CW structure consisting of one 0-cell and k 1-cells.

(i) Show that $d_1^{(2)} : C_1^{(2)}(X) \to C_0^{(2)}(X)$ is surjective.
(ii) Conclude that $b_1^{(2)}(X) = k - 1$ and $b_n^{(2)}(X) = 0$ for $n \neq 1$ and that the reduced and unreduced ℓ^2-homology of X agree.

3.3.2 Let X be a proper, finite type G-CW complex, let I_n be the set of G-orbits of n-cells in X and let $H_i \leq G$ be the stabilizer group of some n-cell in the orbit $i \in I_n$. Show that for each $m \geq 0$ we have the *Morse inequality*

$$\sum_{n=0}^{m} (-1)^{m-n} \sum_{i \in I_n} \frac{1}{|H_i|} \geq \sum_{n=0}^{m} (-1)^{m-n} b_n^{(2)}(X).$$

Explain that the Morse inequalities sharpen both the ℓ^2-Euler Poincaré formula and the *weak Morse inequalities* $\sum_{i \in I_n} \frac{1}{|H_i|} \geq b_n^{(2)}(X)$.

3.3.3 Let G be a group, let $G_0 \leq G$ be a subgroup, and let X be a proper, finite type G_0-CW complex. Show that the G-CW complex $\mathrm{ind}_{G_0}^{G} X$ constructed in Exercise 3.1.2 is likewise proper and finite type and that $b_n^{(2)}(\mathrm{ind}_{G_0}^{G} X) = b_n^{(2)}(X)$ for all $n \geq 0$.

3.4 Cohomological ℓ^2-Betti Numbers

So far we have only dealt with ℓ^2-*homology* arising from the ℓ^2-chain complex of Hilbert modules $C_n^{(2)}(X) = \ell^2 G \otimes_{\mathbb{Z}G} C_*(X)$. It is thus only natural to ask about ℓ^2-*cohomology* which should arise from the "adjoint" cochain complex

$$C_{(2)}^*(X) = \mathrm{Hom}_{\mathbb{Z}G}(C_*(X), \ell^2 G)$$

whose differentials $\delta_{(2)}^*$ are given by precomposing with the cellular differentials d_{*+1}. Similarly as before, each $C_{(2)}^n(X)$ comes with a canonical inner product which can be made explicit by choosing a cellular basis for X. However, $C_{(2)}^n(X)$ is an abelian group of $\mathbb{Z}G$-homomorphisms of *left* $\mathbb{Z}G$-modules. The $\mathbb{Z}G$-$\mathbb{C}G$-bimodule structure on $\ell^2 G$ thus turns $C_{(2)}^n(X)$ into a *right* $\mathbb{C}G$-module. Therefore $(C_{(2)}^*, \delta_{(2)}^*)$ is actually a functor from proper, finite type G-CW complexes to chain complex of right Hilbert $\mathcal{R}(G)$-modules. Here a *right Hilbert $\mathcal{R}(G)$-module* is defined in the only possible way: a Hilbert space H with linear, isometric, right G-action such that there exists a linear, isometric G-embedding $H \hookrightarrow (\ell^2 G)^n$ for some n where we view $(\ell^2 G)^n$ as the diagonal right $\mathbb{C}G$-module. The unit matrix coefficient also gives a trace $\mathrm{tr}_{\mathcal{L}(G)}$ for $\mathcal{L}(G) = \lambda(\mathbb{C}G)'' = B(\ell^2 G)^\rho$ so that we obtain *von Neumann dimension* $\dim_{\mathcal{L}(G)}$ of right Hilbert modules. Given a right Hilbert module H, we can turn it into a left Hilbert module $\mathcal{L}H$ by setting $\mathcal{L}H = H$ as Hilbert spaces but decreeing that $g \in G$ act on $x \in \mathcal{L}H$ by $g \cdot x = xg^{-1}$. Leaving morphisms pointwise unchanged turns $\mathcal{L}$ into a functor $\mathcal{L} \colon \mathrm{mod}\text{-}\mathcal{R}(G) \to \mathcal{L}(G)\text{-mod}$ from right Hilbert modules to left Hilbert modules.

Proposition 3.21 *The functor $\mathcal{L} \colon \mathrm{mod}\text{-}\mathcal{R}(G) \to \mathcal{L}(G)\text{-mod}$ is an equivalence of categories which preserves von Neumann dimension.*

Proof Let $\phi \colon \ell^2 G \to \ell^2 G$ be the flip map $g \mapsto g^{-1}$. We observe that composing a right equivariant embedding $i \colon H \hookrightarrow (\ell^2 G)^n$ with the n-fold product of ϕ yields a left equivariant embedding $\hat{i} \colon \mathcal{L}(H) \hookrightarrow (\ell^2 G)^n$. This shows that $\mathcal{L}$, after all, is a functor. It is clear that building on $x \cdot g = g^{-1}x$ one obtains an inverse functor $\mathcal{R} \colon \mathcal{L}(G)\text{-mod} \to \mathrm{mod}\text{-}\mathcal{R}(G)$. For every direct summand $\ell^2 G$ in $(\ell^2 G)^n$ we compute

$$\langle e, \mathrm{pr}_{i(H)}(e) \rangle = \langle \phi(e), \mathrm{pr}_{i(H)}(e) \rangle = \langle e, \phi \circ \mathrm{pr}_{i(H)} \circ \phi^{-1}(e) \rangle = \langle e, \mathrm{pr}_{\hat{i}(H)}(e) \rangle$$

which gives $\dim_{\mathcal{R}(G)} \mathcal{L}H = \dim_{\mathcal{L}(G)} H$. $\qquad\qquad\square$

We can apply $\mathcal{L}$ to turn the ℓ^2-cochain complex $(C_{(2)}^*(X), \delta_{(2)}^*)$ into a cochain complex of left Hilbert modules. It turns out that this gives the Hilbert space adjoint of the ℓ^2-chain complex.

Proposition 3.22 *Let X be a proper, finite type G-CW complex. Then the cochain complexes $(\mathcal{L}C_{(2)}^n(X), \mathcal{L}\delta_{(2)}^n)$ and $(C_n^{(2)}(X), d_{n+1}^{(2)*})$ of left Hilbert modules are isomorphic.*

Proof As in the preceding section the choice of a cellular basis realizes the cellular differential d_{n+1} by right multiplication

$$
\begin{array}{ccc}
C_{n+1}(X) & \xrightarrow{\;d_{n+1}\;} & C_n(X) \\
\Big\downarrow{\scriptstyle \sim} & & \Big\downarrow{\scriptstyle \sim} \\
\displaystyle\bigoplus_{i\in I_{n+1}} \mathbb{Z}[G/H_i] & \xrightarrow{\;\cdot M\;} & \displaystyle\bigoplus_{j\in I_n} \mathbb{Z}[G/H_j].
\end{array}
$$

with a matrix $M_{ij} \in \mathbb{Z}[h_i\, G/H_j]$. By Propositions 3.8 and 3.9 we obtain that the cellular ℓ^2-differential and its adjoint are given by

$$
\begin{array}{ccc}
C_{n+1}^{(2)}(X) & \xrightarrow{\;d_{n+1}^{(2)}\;} & C_n^{(2)}(X) \\
\Big\downarrow{\scriptstyle \sim} & & \Big\downarrow{\scriptstyle \sim} \\
\displaystyle\bigoplus_{i\in I_{n+1}} \ell^2(G/H_i) & \xrightarrow{\;\cdot M\;} & \displaystyle\bigoplus_{j\in I_n} \ell^2(G/H_j)
\end{array}
\qquad
\begin{array}{ccc}
C_{n+1}^{(2)}(X) & \xleftarrow{\;d_{n+1}^{(2)*}\;} & C_n^{(2)}(X) \\
\Big\downarrow{\scriptstyle \sim} & & \Big\downarrow{\scriptstyle \sim} \\
\displaystyle\bigoplus_{i\in I_{n+1}} \ell^2(G/H_i) & \xleftarrow{\;\cdot M^*\;} & \displaystyle\bigoplus_{j\in I_n} \ell^2(G/H_j)
\end{array}
$$

For the ℓ^2-cochain differential $\delta_{(2)}^n$ the same cellular basis gives a diagram of right Hilbert modules which we turn into a diagram of left Hilbert modules by the equivalence $\mathcal{L}$.

$$
\begin{array}{ccc}
C_{(2)}^{n+1}(X) & \xleftarrow{\;\delta_{(2)}^n\;} & C_{(2)}^n(X) \\
\Big\downarrow{\scriptstyle \sim} & & \Big\downarrow{\scriptstyle \sim} \\
\displaystyle\bigoplus_{i\in I_{n+1}} \ell^2(H_i\backslash G) & \xleftarrow{\;M^\dagger\cdot\;} & \displaystyle\bigoplus_{j\in I_n} \ell^2(H_j\backslash G)
\end{array}
\qquad
\begin{array}{ccc}
\mathcal{L}C_{(2)}^{n+1}(X) & \xleftarrow{\;\mathcal{L}\delta_{(2)}^n\;} & \mathcal{L}C_{(2)}^n(X) \\
\Big\downarrow{\scriptstyle \sim} & & \Big\downarrow{\scriptstyle \sim} \\
\displaystyle\bigoplus_{i\in I_{n+1}} \mathcal{L}\ell^2(H_i\backslash G) & \xleftarrow{\;\mathcal{L}(M^\dagger\cdot)\;} & \displaystyle\bigoplus_{j\in I_n} \mathcal{L}\ell^2(H_j\backslash G)
\end{array}
$$

Here $M^\dagger$ is M viewed as a matrix $M_{ij}^\dagger \in \mathbb{Z}[H_i\backslash G\, h_j]$ which now acts by multiplication from the left and where $\mathcal{L}$ of course preserves colimits. One easily checks that mapping $H_i g \mapsto g^{-1} H_i$ defines an isomorphism of left Hilbert modules $\varphi_i \colon \mathcal{L}\ell^2(H_i\backslash G) \xrightarrow{\sim} \ell^2(G/H_i)$ such that the following square

$$
\begin{array}{ccc}
\displaystyle\bigoplus_{i\in I_{n+1}} \mathcal{L}\ell^2(H_i\backslash G) & \xleftarrow{\;\mathcal{L}(M^\dagger\cdot)\;} & \displaystyle\bigoplus_{j\in I_n} \mathcal{L}\ell^2(H_j\backslash G) \\[2mm]
{\scriptstyle \sim}\Big\downarrow{\scriptstyle \oplus\varphi_i} & & {\scriptstyle \sim}\Big\downarrow{\scriptstyle \oplus\varphi_j} \\[2mm]
\displaystyle\bigoplus_{i\in I_{n+1}} \ell^2(G/H_i) & \xleftarrow{\;\cdot M^*\;} & \displaystyle\bigoplus_{j\in I_n} \ell^2(G/H_j)
\end{array}
$$

commutes. This square connects the two diagrams on the right hand side above which gives the asserted isomorphism of cochain complexes. □

Now let $(C_*^{(2)}, d_*^{(2)})$ be a general chain complex of Hilbert modules. The canonical lift of the reduced homology $H_n^{(2)}(C_*^{(2)}) \hookrightarrow C_n^{(2)}$ has a convenient description as the kernel of the n-th ℓ^2-Laplacian $\Delta_n^{(2)} = d_n^{(2)*} d_n^{(2)} + d_{n+1}^{(2)} d_{n+1}^{(2)*}$ as we show next.

Proposition 3.23 *We have the ℓ^2-Hodge–de-Rham decomposition*

$$C_n^{(2)} = \ker \Delta_n^{(2)} \oplus \overline{\operatorname{im} d_{n+1}^{(2)}} \oplus \overline{\operatorname{im} d_n^{(2)*}}$$

and the canonical map $\ker \Delta_n^{(2)} \to H_n^{(2)}(C_*^{(2)})$ *is an isomorphism for* $n \geq 0$.

Proof In view of the orthogonal decomposition $C_n^{(2)} = \ker d_n^{(2)} \oplus \overline{\operatorname{im} d_n^{(2)*}}$ it only remains to show that $\ker \Delta_n^{(2)} = \ker d_n^{(2)} \cap \ker d_{n+1}^{(2)*}$ because the right hand side is the orthogonal complement of $\overline{\operatorname{im} d_{n+1}^{(2)}}$ in $\ker d_n^{(2)}$ and thus the canonical lift of $H_n^{(2)}(C_*^{(2)})$. But this identity follows from

$$\langle \Delta_n^{(2)} x, x \rangle = \left\| d_n^{(2)} x \right\|^2 + \left\| d_{n+1}^{(2)*} x \right\|^2.$$ □

The proposition can easily be transferred to cochain complexes and the cochain complex $(C_n^{(2)}, d_{n+1}^{(2)*})$ and its adjoint chain complex $(C_n^{(2)}, d_n^{(2)})$ share the same ℓ^2-Laplacian. Consequently, the reduced homology of $(C_n^{(2)}, d_n^{(2)})$ is equal to the reduced cohomology of $(C_n^{(2)}, d_{n+1}^{(2)*})$. Therefore the last three propositions combine to the following theorem about the n-th *reduced ℓ^2-cohomology* $H_{(2)}^n(X) = \ker \delta_{(2)}^n / \overline{\operatorname{im} \delta_{(2)}^{n-1}}$ and the *cohomological n-th ℓ^2-Betti number* $b_{(2)}^n(X) = \dim_{\mathcal{L}(G)} H_{(2)}^n(X)$.

Theorem 3.24 *Let X be a proper, finite type G-CW complex. Then for all $n \geq 0$ we have* $H_n^{(2)}(X) \cong \mathcal{L}H_{(2)}^n(X)$ *and* $b_n^{(2)}(X) = b_{(2)}^n(X)$.

We are now prepared to prove that one of the cornerstones of algebraic topology, *Poincaré duality*, has an ℓ^2-counterpart.

Theorem 3.25 (ℓ^2-Poincaré Duality) *Let M be an orientable G-manifold of dimension m which comes with a triangulation as a finite, free G-CW complex. Then* $b_n^{(2)}(G \curvearrowright M) = b_{m-n}^{(2)}(G \curvearrowright M)$.

Proof While in general a free G-space X must not be proper (in the sense that the graph map $G \times X \to X \times X$ is proper), a free G-CW complex always is. Therefore the quotient map $M \to G \backslash M$ is a Galois covering and $G \backslash M$ is a closed manifold. We have a homomorphism $G \to \mathbb{Z}/2\mathbb{Z}$ which sends g to 0 or 1 according to whether g acts orientation preserving or reversing on M. Thus G has a subgroup of index at most two acting orientation preservingly.

By the restriction property, Theorem 3.18 (iv), we may assume G itself acts orientation preservingly which implies that $G \backslash M$ is orientable. According to [169, Theorem 2.1, p. 23] this guarantees the existence of a $\mathbb{Z}G$-chain homotopy equivalence

$$[G \backslash M] \cap (\cdot) \colon \mathcal{L}C^{m-*}(M) \to C_*(M).$$

Here the left hand side is the chain complex (chain, not cochain!) given by $C^{m-*}(M) = \operatorname{Hom}_{\mathbb{Z}G}(C_{m-*}(M), \mathbb{Z}G)$ which we turned into a chain complex of left $\mathbb{Z}G$-modules by reverting the $\mathbb{Z}G$-action by means of the canonical involution of $\mathbb{Z}G$. We have allowed ourselves to denote the latter process by $\mathcal{L}$ just like in the case of Hilbert modules. Actually, the above reference provides this chain homotopy only for universal coverings and under the convention that deck transformation is a right action but neither of this is problematic. As in the proof of Theorem 3.18 (i) the chain homotopy equivalence gives an isomorphism of reduced homology after going over to the ℓ^2-completions. Moreover, the chain complexes of left Hilbert modules $\ell^2 G \otimes_{\mathbb{Z}G} \mathcal{L}C^{m-*}(M)$ and $\mathcal{L}C_{(2)}^{m-*}(M)$ are naturally isomorphic by $x \otimes f \mapsto (y \mapsto f(y)x^*)$ where "$*$" denotes the unitary involution of $\ell^2 G$ given by $g \mapsto g^{-1}$. Together with Theorem 3.24 we conclude $H_n^{(2)}(M) \cong H_{m-n}^{(2)}(M)$. □

Of course, choosing a different triangulation for M does not alter the ℓ^2-Betti numbers by homotopy invariance, Theorem 3.18 (i). Equivariant triangulations for smooth, proper G-manifolds are known to exist so that ℓ^2-Poincaré duality can also be given as a mere statement on manifolds.

To not overload the presentation notationally we have withheld so far the information that all occurring (co)chain complexes have relative versions coming from G-CW pairs (X, A). These give rise to *relative ℓ^2-Betti numbers* $b_n^{(2)}(X, A)$. With no additional effort, also Poincaré–Lefschetz duality extends to the ℓ^2-setting and gives the more general *ℓ^2-Poincaré–Lefschetz formula* $b_n^{(2)}(M) = b_{m-n}^{(2)}(M, \partial M)$ for smooth free, proper, cocompact, m-dimensional G-manifolds with (empty or nonempty) boundary.

Example 3.26 Let Σ_g be the closed, orientable surface of genus $g \geq 1$. Then $b_0^{(2)}(\widetilde{\Sigma}_g) = 0$ by Theorem 3.18 (ii) because Σ_g is connected and $\pi_1(\Sigma_g)$ is infinite. Thus ℓ^2-Poincaré duality gives $b_2^{(2)}(\widetilde{\Sigma}_g) = 0$. By the ℓ^2-Euler-Poincaré formula, Theorem 3.19, we must then have $b_1^{(2)}(\widetilde{\Sigma}_g) = -\chi(\Sigma_g) = 2g - 2$. Since there are no cells of dimension three or higher, all other ℓ^2-Betti numbers are zero.

Now let us remove $d \geq 1$ open disks from Σ_g to obtain the surface of genus g with d punctures $\Sigma_{g,d}$. Figure 3.3 explains that $\Sigma_{g,d}$ deformation retracts to a wedge sum of $2g + d - 1$ circles, $\Sigma_{g,d} \simeq \bigvee_{i=1}^{2g+d-1} S^1$. Thus $\pi_1(\Sigma_{g,d})$ is free on $(2g + d - 1)$ letters and hence $b_0^{(2)}(\widetilde{\Sigma}_{g,d}) = 0$. By homotopy invariance, Theorem 3.18 (i), and by the ℓ^2-Euler-Poincaré formula we obtain $b_1^{(2)}(\widetilde{\Sigma}_{g,d}) = d + 2(g - 1)$.

Fig. 3.3 The surface of
genus g with d punctures
deformation retracts to a
wedge sum of $2g + d - 1$
circles. The sides of the
$4g$-gon are identified in pairs.
Additionally, we see that we
obtain one circle less than the
number of punctures

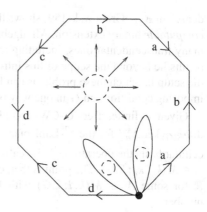

By Poincaré duality we have $b_0^{(2)}(\widetilde{\Sigma}_{g,d}, \partial\widetilde{\Sigma}_{g,d}) = b_2^{(2)}(\widetilde{\Sigma}_{g,d}, \partial\widetilde{\Sigma}_{g,d}) = 0$ and $b_1^{(2)}(\widetilde{\Sigma}_{g,d}, \partial\widetilde{\Sigma}_{g,d}) = d + 2(g - 1)$.

3.5 Atiyah's Question and Kaplansky's Conjecture

What are possible values of ℓ^2-Betti numbers? By definition ℓ^2-Betti numbers of G-CW complexes are nonnegative real numbers. It turns out, however, that all ℓ^2-Betti numbers we have computed so far are actually rational numbers. Atiyah made a similar observation when he originally introduced ℓ^2-Betti numbers in an analytic context and asked for examples of irrational ℓ^2-Betti numbers [8, p. 72]. Translated to our setting the question takes the following form.

Question 3.27 (Atiyah [8]) Does there exist a finite, connected, free G-CW complex X such that $b_n^{(2)}(G \curvearrowright X) \notin \mathbb{Q}$ for some $n \geq 0$?

Observe that assuming X is connected requires G to be finitely generated. The question remained open for 30-some-odd years until Tim Austin [10] answered it in the least constructive way one can imagine.

Theorem 3.28 (Austin [10]) *Let $B^{(2)} \subset \mathbb{R}_{\geq 0}$ be the set of real numbers which occur as an ℓ^2-Betti number of a finite, connected, free G-CW complex X. Then $B^{(2)}$ is uncountable.*

In particular, $B^{(2)}$ contains irrational and even transcendental elements. Shortly thereafter, Grabowski [59] and Pichot–Schick–Zuk [145] showed independently that in fact $B^{(2)} = \mathbb{R}_{\geq 0}$. Also of interest is the subset $\widetilde{B}^{(2)}$ where one additionally requires that X be simply-connected. By what we said in Example 3.4 the set $\widetilde{B}^{(2)}$ is precisely the set of all ℓ^2-Betti numbers of universal coverings of finite connected CW complexes. Since there are only countably many finite CW complexes up to homotopy equivalence, it follows that $\widetilde{B}^{(2)}$ is countable. But $\widetilde{B}^{(2)}$ seems to be "very dense" in $\mathbb{R}_{\geq 0}$. Grabowski [59] shows that it contains all nonnegative numbers with *computable* binary extension. All algebraic numbers fall into this category and also many transcendental ones, including π and e. The proof methods of these recent results lie beyond the scope of this introductory course. But we want to understand the setup in which these problems can be attacked as it also allows positive results, imposing conditions on G in one way or another.

Given a finite, free G-CW complex X, Proposition 3.23 says $b_n^{(2)}(X) = \dim_{\mathcal{R}(G)} \ker \Delta_n^{(2)}$ for the ℓ^2-Laplacian $\Delta_n^{(2)} = d_n^{(2)*} d_n^{(2)} + d_{n+1}^{(2)} d_{n+1}^{(2)*}$. With a fixed cellular basis the operator $\Delta_n^{(2)}$ is realized by right multiplication with some matrix $A \in M(k_n, k_n; \mathbb{Z}G)$. The point is that, conversely, every number $\dim_{\mathcal{R}(G)} \ker(\cdot A) \in \mathbb{R}$ for some $A \in M(k, l; \mathbb{Z}G)$ with finitely generated G occurs as an ℓ^2-Betti number.

Proposition 3.29 *Let $A \in M(k, l; \mathbb{Z}G)$ be a matrix over the integral group ring of a group G that is generated by r elements. Then there exists a free G-CW complex X consisting of k equivariant 3-cells, l equivariant 2-cells, r equivariant 1-cells and one equivariant 0-cell such that the third ℓ^2-differential $d_3^{(2)} : C_3^{(2)}(X) \longrightarrow C_2^{(2)}(X)$ can be identified with the right multiplication operator $(\ell^2 G)^k \xrightarrow{\cdot A} (\ell^2 G)^l$.*

Proof For simplicity, we start with the case $k = l = 1$. Consider the space $Z = S^2 \vee (\bigvee_{j=1}^r S^1)$. We attach a 3-cell to Z by an attaching map $\varphi : S^2 \to Z$ described as follows. Write the only entry of A as $a = \sum_{s=1}^N a_s w_s(g_1, \ldots, g_r)$ where the w_s are words in the generators $g_1, \ldots, g_r \in G$. Embed N little open 2-disks into S^2. Collapsing the complement of these disks to a point we obtain a wedge sum of N little 2-spheres. Say the common base point of these 2-spheres is the south pole in each of the 2-spheres. Then we collapse all circles of latitude in the southern hemispheres up to the equator to one point each. The resulting space looks like a bunch of lollipops stuck together at the free ends as illustrated in Fig. 3.4. From this space the element $a \in \mathbb{Z}G$ determines the map to Z: We give an orientation to the 1-cells of Z and label them by the generators $g_1, \ldots, g_r \in G$. Now the base point goes to the base point, the stick of the s-th lollipop is wrapped around the r one-cells of Z according to the word w_s and the candy 2-sphere of the s-th lollipop is mapped to the only copy of S^2 in Z by a map of degree a_s. Set $Y = D^3 \cup_\varphi Z$. The labeling of the 1-cells in Y determines an epimorphism $\pi_1 Y \to G$ where $\pi_1 Y$ is free of rank k. The corresponding Galois covering of Y with deck transformation group G is the desired G-CW complex X. Indeed, the characteristic map $\Phi : D^3 \to Y$ of the 3-cell lifts to G-many characteristic maps $D^3 \to X$. After choosing a base point in the 0-skeleton of X these form the characteristic map $Q : G \times D^3 \to X$ of an equivariant 3-cell and the cellular differential of the cell $Q(\{e\} \times D^3)$ is by construction the element $a \in \mathbb{Z}G \cong C_2(X)$.

The adaptation to the general case is easy. We start with $Z = (\bigvee_{a=1}^l S^2) \vee (\bigvee_{b=1}^r S^1)$ and for each $i = 1, \ldots, k$ we attach one 3-cell as follows. We embed l families of 2-disks into S^2 corresponding to the entries of the i-th row of A. Then we collapse as above and the entry A_{ij} determines how the j-th bunch of lollipop

Fig. 3.4 The quotient space
of S^2 after the two collapsing
procedures. The base point
corresponds to the
complement of the embedded
disks and the sticks of the
lollipops are formed by
collapsed circles of latitude

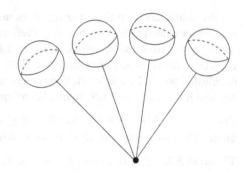

is mapped to the subspace of Z consisting of the one-skeleton and the j-th copy of
S^2. If Y denotes Z with the k 3-cells attached, then again the Galois covering X
corresponding to the epimorphism $\pi_1 Y \to G$ does the trick.   □

If G is finitely presented, we can additionally attach finitely many 2-cells to the
CW complex Y from the above proof corresponding to the finitely many relators of
G. The universal covering is a simply-connected G-CW complex X whose third ℓ^2-
differential is given by right multiplication with $A \times 0$ where the nil-factor means that
the differential always assigns zero coefficients to the relator cells. Thus identifying
the sets $B^{(2)}$ and $\widetilde{B}^{(2)}$ has become a mere problem in operator algebras and group
theory. In fact, for a ring $\mathbb{Z} \subseteq R \subseteq \mathbb{C}$ let

$$B_R^{(2)}(G) = \{\dim_{\mathcal{R}(G)} \ker(\cdot A) \colon A \in M(k, l; RG); k, l \geq 1\},$$

then as a consequence of Proposition 3.29 and our discussion we have

$$B^{(2)} = \bigcup_{G \text{ f.g.}} B_{\mathbb{Z}}^{(2)}(G) \qquad \text{and} \qquad \widetilde{B}^{(2)} = \bigcup_{G \text{ f.p.}} B_{\mathbb{Z}}^{(2)}(G)$$

forming the union over all finitely generated groups and all finitely presented groups,
respectively.

Actually, all of the groups considered by Austin, Grabowski and Pichot–Schick–
Zuk which lead to examples of irrational ℓ^2-Betti numbers arise as wreath products
of infinite and finite groups and thus possess arbitrarily large finite subgroups. This
property seems to be essential in the arguments. So we might suspect that groups
with an upper bound on the order of finite subgroups cannot lead to irrational ℓ^2-
Betti numbers. In fact we should be more specific because in the few example
computations of ℓ^2-Betti numbers we have done so far, we could only produce non-
integer ℓ^2-Betti number whose denominators were given by the order of stabilizer
subgroups.

Conjecture 3.30 (Atiyah Conjecture)   Let $\mathbb{Z} \subseteq R \subseteq \mathbb{C}$ be a ring and let G be a
group whose finite subgroups are of bounded order. Denote by $\mathrm{lcm}(G)$ the least
common multiple of all occurring orders. Then $B_R^{(2)}(G) \subseteq \frac{1}{\mathrm{lcm}(G)}\mathbb{Z}_{\geq 0}$.

The conjecture with these particular assumptions is due to Lück and Schick. It is however named after Atiyah, being a refinement of his original question 3.27. To distinguish it from this question, it is sometimes also known as the *strong* Atiyah conjecture for G, especially when $R = \mathbb{C}$. Note that for a torsion-free group G, the conclusion would be that ℓ^2-Betti numbers are integers. This is worth mentioning because it has a serious mathematical consequence.

Conjecture 3.31 (Kaplansky) Let $\mathbb{Z} \subseteq R \subseteq \mathbb{C}$ be a ring and let G be a torsion-free group. Then the group ring RG has no nontrivial zero divisors.

Theorem 3.32 *If G is torsion-free and satisfies the Atiyah conjecture with coefficients in R, then RG satisfies the Kaplansky conjecture.*

Proof Suppose $a, b \in RG$ satisfy $a \cdot b = 0$. If $a = 0$, we are done. Otherwise, $a \in RG \subset \ell^2 G$ is a nonzero element in the kernel of the Hilbert module morphism $\ell^2 G \xrightarrow{\cdot b} \ell^2 G$. Thus $0 < \dim_{\mathcal{R}(G)} \ker(\cdot b) \le 1$ by the faithfulness and normalization properties of von Neumann dimension, Theorem 2.44 (i) and (ii). Since we assume the Atiyah conjecture for R and G, we conclude $\dim_{\mathcal{R}(G)} \ker(\cdot b) = 1$. Of course we have $\ell^2 G = \ker(\cdot b) \oplus (\ker(\cdot b))^{\perp}$ so that by additivity, normalization, and faithfulness (Theorem 2.44), we have $\ker(\cdot b) = \ell^2 G$. In particular, for the unit element vector $e \in \ell^2 G$ this gives $b = e \cdot b = 0$. $\square$

This also settles Theorem 1.2 from the introduction as we discuss next. It should be apparent by now that the somewhat clumsy assumptions of this theorem are in place to account for a possibly infinitely generated group G.

Proof (of Theorem 1.2) Suppose $a, b \in \mathbb{Q}G$ satisfy $a \cdot b = 0$. Since both a and b are finite linear combinations of elements in G, we have $a, b \in \mathbb{Q}H$ for some finitely generated subgroup $H \le G$ and H is of course still torsion-free. The assumption of the theorem assures that $B_{\mathbb{Z}}^{(2)}(H) \subseteq \mathbb{Z}_{\ge 0}$ via Proposition 3.29. This implies $B_{\mathbb{Q}}^{(2)}(H) \subseteq \mathbb{Z}_{\ge 0}$ because clearing denominators leaves the kernels unchanged. The last theorem thus gives $a = 0$ or $b = 0$. $\square$

You will prove in Exercise 3.5.1 that the Kaplansky conjecture 3.31 on zero divisors also implies that RG has neither nilpotent nor idempotent elements. This provides even more motivation for finding positive results on the Atiyah conjecture. Before we can state such a result, we have to give some definitions. A group is called *elementary amenable* if it belongs to the smallest class of groups $\mathcal{E}$ such that $\mathcal{E}$ is closed under taking subgroups, quotients, extensions, and directed unions and such that $\mathcal{E}$ contains all finite and abelian groups. One can think of $\mathcal{E}$ as those groups that are not "too far" from being virtually abelian. For example, virtually solvable groups are elementary amenable. Let moreover C be the smallest class of groups that is closed under directed unions and contains all free groups and all groups G which occur in an extension

$$1 \longrightarrow N \longrightarrow G \longrightarrow A \longrightarrow 1$$

with $N \in C$ and A elementary amenable. The first substantial result on the Atiyah conjecture is due to P. Linnell [105].

Theorem 3.33 (Linnell [105]) *Suppose the finite subgroups of $G \in C$ have bounded order. Then G satisfies the Atiyah conjecture with $R = \mathbb{C}$.*

The proof of this theorem is an impressive compound of noncommutative algebra and operator algebra theory that goes beyond what could be included here. Nonetheless, T. Schick had the fruitful idea that approximation techniques can substantially extend the class of groups for which the Atiyah conjecture is known, at least if $R = \mathbb{Q}$. We will come back to this point in Sect. 5.6 of Chap. 5 where we will exemplify this trick by showing the Atiyah conjecture with $R = \mathbb{Q}$ for free groups if one only accepts it for torsion-free elementary amenable groups. We will then move on to discuss further extensions and survey how a vast generalization of Theorem 3.33 has emerged recently as Theorem 5.58.

Exercises

3.5.1 Let R be an integral domain and let G be a torsion-free group. We say that the group ring RG satisfies the *Kaplansky conjecture on*

  (i) *units* if every unit in RG is of the form rg for some $r \in R^*$ and $g \in G$,
 (ii) *nilpotents* if every nilpotent element in RG is trivial or, equivalently, $a^2 = 0$ implies $a = 0$ in RG,
(iii) *zero divisors* if every zero-divisor in RG is trivial,
(iv) *idempotents* if every idempotent in RG is trivial: if $a^2 = a$ in RG, then $a = 0$ or $a = 1$.

Show the implications (i) $\Rightarrow$ (ii) $\Leftarrow$ (iii) $\Rightarrow$ (iv). *Remark: Actually, it is also known that* (ii) $\Rightarrow$ (iii) *but the proof is somewhat challenging. Of course, no counterexamples to the remaining two implications are known, since all conjectures might be true.*

3.5.2 Let $\beta_0, \beta_1, \ldots, \beta_N$ be nonnegative rational numbers. Find a group G and a finite, proper G-CW complex X with $b_n^{(2)}(X) = \beta_n$ for $n = 0, 1, \ldots, N$.

3.6 ℓ^2-Betti Numbers as Obstructions

Frequently one wants to prove that some mathematical object does not admit a certain additional structure. To do so, one should find a nonzero *obstruction*: an invariant which vanishes for all objects possessing the additional structure but does not vanish for the object under investigation. Famous examples in topology are the *Stiefel–Whitney numbers* which obstruct that a closed smooth n-manifold is the boundary of an $(n + 1)$-manifold, or the $\hat{A}$-*genus* which obstructs the existence of a positive scalar curvature metric on a $4k$-dimensional spin manifold. In this section we will see that ℓ^2-Betti numbers can also be interpreted as obstructions in various contexts.

3.6.1 ℓ^2-Betti Numbers Obstruct Nontrivial Self-coverings

The classical Euler characteristic $\chi(X)$ obstructs the existence of non-trivial self-coverings of a connected CW complex X. This is immediate from the multiplicative behavior under finite coverings: if $X \to Y$ is a d-sheeted covering, then $\chi(X) = d \cdot \chi(Y)$. Even though $\chi(X)$ equals the alternating sum of ordinary Betti numbers, the latter are not multiplicative individually as the twofold self-covering of the circle reveals. Thus ordinary Betti numbers are not much good for deciding about the existence of self-coverings. But according to Theorem 3.19, $\chi(X)$ is also the alternating sum of ℓ^2-Betti numbers and it turns out that these are multiplicative individually.

Proposition 3.34 *Let* $X \xrightarrow{p} Y$ *be a* d-*sheeted covering of connected CW complexes of finite type. Then* $b_n^{(2)}(\widetilde{X}) = d \cdot b_n^{(2)}(\widetilde{Y})$ *for all* $n \geq 0$.

Proof We have a tower of coverings $\widetilde{X} \to X \xrightarrow{p} Y$. An element $y \in \pi_1 Y$ acts on $\widetilde{X}$ as a deck transformation of X if and only if y lies in the image of $\pi_1 X \xrightarrow{p_*} \pi_1 Y$. So while $\widetilde{X}$ and $\widetilde{Y}$ are equal as CW complexes, as G-CW complexes we have $\widetilde{X} = \mathrm{res}^{\pi_1 Y}_{p_* \pi_1 X} \widetilde{Y}$. From Theorem 3.18 (iv) it follows that $b_n^{(2)}(\widetilde{X}) = [\pi_1 Y : p_* \pi_1 X] \, b_n^{(2)}(\widetilde{Y}) = d \cdot b_n^{(2)}(\widetilde{Y})$.   □

Corollary 3.35 *Let* X *be a connected, finite type CW complex. If* $b_n^{(2)}(\widetilde{X}) > 0$ *for some* $n \geq 0$, *then* X *does not have any nontrivial, connected, finite sheeted self coverings.*

For example the k-dimensional torus $\mathbb{T}^k$ has many self coverings, thus $b_n^{(2)}(\widetilde{\mathbb{T}^k}) = 0$ for all $n \geq 0$ as could equally well be deduced from the Künneth formula. Note that ℓ^2-Betti numbers provide an a priori sharper obstruction to self-coverings than the Euler characteristic: if all ℓ^2-Betti numbers vanish, then so does the Euler characteristic but the converse is wrong.

3.6.2 ℓ^2-Betti Numbers Obstruct Mapping Torus Structures

For many spaces, and in particular for 3-manifolds, it is an intriguing question if a space can be constructed as a *mapping torus*. Let us first recall the definition.

Definition 3.36 Let X be a topological space and let $f: X \to X$ be a (continuous) map. The *mapping torus* $T(f)$ is the quotient space of $X \times I$ obtained by identifying $(x, 1)$ with $(f(x), 0)$ for all $x \in X$.

For example, let $X = S^1$ be the circle. If f is the identity map, we have $T(f) = \mathbb{T}^2$, whereas if f is complex conjugation, the mapping torus $T(f) = K$ is the Klein bottle. We observe that $T(f)$ always comes with a canonical map $T(f) \xrightarrow{p} S^1$ sending (x, t) to $e^{2\pi i t}$.

Lemma 3.37 *If X is path-connected, then $\pi_1(p)$ is surjective.*

Proof Fix a base point $x_0 \in X$ and a path γ from x_o to $f(x_0)$. Then the loop $\overline{\gamma}: I \to T(f)$ defined by $t \mapsto [(\gamma(t), t)]$ maps to a generator of $\pi_1(S^1)$ under $\pi_1(p)$. □

Theorem 3.38 (Lück, [114]) *Let X be a connected, finite type CW complex and let $f: X \to X$ be cellular. Then $b_n^{(2)}(\widetilde{T(f)}) = 0$ for all $n \geq 0$*

Proof Let us set $G = \pi_1(T(f))$. For each $k \geq 1$ we consider the subgroup $G_k = \pi_1(p)^{-1}(k\mathbb{Z})$ of G. The k-fold covering space $\overline{T(f)}_{G_k}$ of $T(f)$ corresponding to G_k can be constructed by gluing k copies of the product $X \times I$ cyclically, always along the map f, as suggested by Fig. 3.5. Retracting all but one copy of $X \times I$ along the I-coordinate defines a homotopy equivalence from $\overline{T(f)}_{G_k}$ to $T(f^k)$. Since $\widetilde{\overline{T(f)}_{G_k}} \cong \mathrm{res}_{G_k}^G \widetilde{T(f)}$, we obtain

$$b_n^{(2)}(\widetilde{T(f)}) = \frac{b_n^{(2)}(\mathrm{res}_{G_k}^G(\widetilde{T(f)}))}{k} = \frac{b_n^{(2)}(\widetilde{\overline{T(f)}_{G_k}})}{k} = \frac{b_n^{(2)}(\widetilde{T(f^k)})}{k}$$

from Theorem 3.18 (iv) and (i). Clearly if X has c_n n-cells, then $T(f^k)$ has $c_n + c_{n-1}$ many n-cells so that $b_n^{(2)}(\widetilde{T(f)}) \leq \frac{c_n + c_{n-1}}{k}$ by the weak Morse inequality from Exercise 3.3.2. Since this holds for any k, it follows that $b_n^{(2)}(\widetilde{T(f)}) = 0$ for all $n \geq 0$. □

In Sect. 6.4 of Chap. 6, we will report on a recent break through in 3-manifolds. To wit, Theorem 6.19 implies that each closed hyperbolic 3-manifold has a finite covering which is a mapping torus $T(f)$ of a homeomorphism $f: \Sigma_g \to \Sigma_g$ for some surface Σ_g. Proposition 3.34 and Theorem 3.38 therefore yield the following result.

Theorem 3.39 *A closed hyperbolic 3-manifold is ℓ^2-acyclic.*

Fig. 3.5 Schematic picture
of the finite cyclic covering of
a mapping torus which arises
by cyclically gluing copies of
the cylinder $X \times I$ by means
of the map f

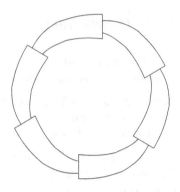

More precisely, the universal covering of such a manifold is ℓ^2-acyclic but statements of this kind are customary and unlikely lead to confusion. As Theorem 6.19 reveals, Theorem 3.39 actually holds for a way more general class of 3-manifolds. But it should be contrasted with the case of hyperbolic surfaces which have a nonzero first ℓ^2-Betti number as we have seen in Example 3.26. We remark that long before Theorem 6.19 was available, Dodziuk [35] showed that all odd-dimensional closed hyperbolic manifolds are ℓ^2-acyclic whereas the even dimensional ones have a positive ℓ^2-Betti number precisely in the middle dimension. He showed this in the analytic approach to ℓ^2-Betti numbers via de Rham cohomology of square integrable differential forms on Riemannian manifolds with isometric, cocompact G-action. It is actually in the latter setting that ℓ^2-Betti numbers were originally defined by Atiyah [8]. These analytically defined ℓ^2-Betti numbers equal our cellular ℓ^2-Betti numbers for any G-CW structure. The proof is likewise due to Dodziuk [34].

Mapping tori of self-homeomorphisms of some space X are also known as *fiber bundles* over S^1 with fiber X. In general, a fiber bundle over a base space B with fiber X can be viewed locally as a product of X and B but globally, the space might be twisted; see for instance [68, Section 4.2]. Mapping tori of self-homotopy equivalences fall under the weaker concept of *fibrations* over S^1, compare Exercise 4.5.1. Conversely, every fibration over S^1 arises up to homotopy equivalence as a mapping torus. Hence, ℓ^2-Betti numbers obstruct that a space has the structure of a fibration (let alone a fiber bundle) over the circle.

3.6.3 ℓ^2-Betti Numbers Obstruct Circle Actions

Recall from below Definition 3.3 that we defined G-CW complexes also for a possibly non-discrete topological group G such as the circle group S^1 of unit complex numbers.

Theorem 3.40 *Let X be a connected, finite type S^1-CW complex such that some S^1-orbit of X embeds π_1-injectively into X. Then every S^1-orbit embeds π_1-injectively into X and $b_n^{(2)}(\widetilde{X}) = 0$ for all $n \geq 0$.*

Proof Let $x_0 \in X$ be a point in the π_1-injectively included orbit and let $y_0 \in X$ be any point outside this orbit. Since X is a connected CW complex, it is also path connected, and we can pick a path γ from x_0 to y_0. Traveling along the path to some point of the path, then traveling through the S^1-orbit of this point and finally traveling the path backwards defines a homotopy between the loop in X based at x_0 given by the S^1-orbit of x_0 and the loop $\gamma \cdot \varphi \cdot \gamma^{-1}$ where φ is the loop based at y_0 given by the S^1-orbit of y_0. This shows that all orbits include π_1-injectively. For the second statement, we show more generally that the pullback $f^*\widetilde{X}$ of the universal covering $\widetilde{X} \to X$ along any S^1-equivariant, cellular map $f: Y \to X$ from a finite type S^1-CW complex Y is ℓ^2-acyclic as $\pi_1 X$-CW complex. The theorem then follows by setting $f = \mathrm{id}_X$. Since $b_n^{(2)}(\widetilde{Y})$ depends on the $(n+1)$-skeleton only, we can assume that Y is finite and prove the theorem by induction on the dimension N. For $N = 0$ the statement is vacuous (and thus true). Let Y be N-dimensional and consider the S^1-equivariant pushout that produces Y from the $(N-1)$-skeleton Y_{N-1}.

$$
\begin{array}{ccc}
\coprod_{i \in I_N} S^1/H_i \times S^{N-1} & \xrightarrow{\coprod q_i} & Y_{N-1} \\
\downarrow{\scriptstyle i} & & \downarrow{\scriptstyle j} \\
\coprod_{i \in I_N} S^1/H_i \times D^N & \xrightarrow{\coprod Q_i} & Y
\end{array}
$$

Pulling back the universal covering $\widetilde{X} \to X$ along f and its precompositions with the maps of the pushout diagram yields a $\pi_1 X$-equivariant pushout

$$
\begin{array}{ccc}
\coprod_{i \in I_N} (j \circ q_i)^* f^*\widetilde{X} & \longrightarrow & j^* f^*\widetilde{X} \\
\downarrow & & \downarrow \\
\coprod_{i \in I_N} Q_i^* f^*\widetilde{X} & \longrightarrow & f^*\widetilde{X}.
\end{array}
$$

Recall from [85, Theorem 5.15, p. 56] that a (non-equivariant) pushout diagram gives rise to a long exact sequence of ordinary homology groups, called the *Mayer–Vietoris sequence*. Along similar lines one can show that an equivariant pushout gives rise to a long *weakly* exact sequence in ℓ^2-homology, though we skip the somewhat tedious proof which requires checking many details [117, Theorem 1.21,

p. 27]. For the above pushout the sequence looks like

$$\cdots \to \bigoplus_{i \in I_n} H_*^{(2)}((j \circ q_i)^* f^* \widetilde{X}) \to H_*^{(2)}(j^* f^* \widetilde{X}) \oplus \bigoplus_{i \in I_n} H_*^{(2)}(Q_i^* f^* \widetilde{X}) \to$$

$$\to H_*^{(2)}(f^* \widetilde{X}) \to \bigoplus_{i \in I_n} H_{*-1}^{(2)}((j \circ q_i)^* f^* \widetilde{X}) \to \cdots .$$

The S^1-CW complexes $S^1/H_i \times S^{N-1}$ and Y_{N-1} are of lower dimension so that the $\pi_1 X$-CW complexes $(j \circ q_i)^* f^* \widetilde{X}$ and $j^* f^* \widetilde{X}$ are ℓ^2-acyclic by induction hypothesis. It thus remains to show that the $\pi_1 X$-CW complexes $Q_i^* f^* \widetilde{X}$ are ℓ^2-acyclic. The space $S^1/H_i \times D^N$ is homotopy equivalent to S^1 because the S^1-action cannot have fixed points as this would violate the π_1-injectivity condition. Therefore $Q_i^* f^* \widetilde{X}$ is $\pi_1 X$-homotopy equivalent to a (generally non-connected) covering of S^1 which must be of the form $\pi_1 X \times_{\mathbb{Z}} \widetilde{S^1}$ for an embedding $\mathbb{Z} \hookrightarrow \pi_1 X$ because S^1/H_i includes π_1-injectively into X by assumption. Here for $H \leq G$, the notation $G \times_H Z$ denotes the G-CW complex *induced* from the H-CW complex Z constructed in Exercise 3.1.2. Concretely, it can be defined by identifying (g, z) with (gh^{-1}, hz) for all $h \in H$ in the product $G \times Z$. By Exercise 3.3.3 ℓ^2-Betti numbers remain unchanged under induction so that $Q_i^* f^* \widetilde{X}$ is ℓ^2-acyclic because S^1 is, as we had already seen in Example 3.15. □

As a consequence we obtain the first half of Theorem 1.3 from the introduction.

Corollary 3.41 *An even dimensional closed hyperbolic manifold M does not permit any nontrivial action by the circle group.*

Proof To conclude the corollary from Theorem 3.40, we still have to apply a couple of nontrivial results which we take for granted as they lie outside the field of ℓ^2-invariants. First of all, associated with any S^1-action on M we have a finite S^1-CW structure according to an equivariant triangulation theorem due to Illman [78, Corollary 7.2]. Next one proves that a nontrivial S^1-action on an aspherical manifold cannot have fixed points. Afterwards one verifies that a finite S^1-CW complex without fixed points satisfying $\widetilde{H}_*(\widetilde{X}; \mathbb{Q}) = 0$ has π_1-injectively embedded orbits; see [117, Lemma 1.42, p.45] for both statements. Thus if M had a nontrivial S^1-action, it would satisfy the assumptions of Theorem 3.40. The conclusion of the theorem contradicts Dodziuk's result mentioned above that an even dimensional closed hyperbolic manifold has a nonzero middle ℓ^2-Betti number. □

The $\hat{A}$-*genus* of a compact oriented spin $4k$-manifold [9], Gromov's *simplicial volume* of an oriented closed connected manifold [62, 109], and ℓ^2-torsion, to be studied in Chap. 6, are all likewise obstructions to finite self-coverings and circle actions. But neither simplicial volume nor ℓ^2-torsion vanishes for mapping tori because for a hyperbolic 3-manifold, both are proportional to the volume with a positive constant.

Chapter 4
ℓ^2-Betti Numbers of Groups

A powerful outcome of topology is the fact that any homotopy invariant of spaces yields an isomorphism invariant of groups via the construction of *classifying spaces*. We introduce this concept right away in the version relative to *families of subgroups*. This turns out to be useful because we defined ℓ^2-Betti numbers not only for free but also for proper G-CW complexes.

4.1 Classifying Spaces for Families

Definition 4.1 A *family of subgroups* $\mathcal{F}$ of G is a set of subgroups of G which is closed under conjugation and finite intersections.

Examples are given by the trivial family $\mathcal{TRIV}$ consisting only of the trivial subgroup, the family $\mathcal{ALL}$ of all subgroups, the family of finite subgroups $\mathcal{FIN}$ and the family $\mathcal{VCYC}$ of *virtually* cyclic subgroups. Here we apply the meta definition that a group has *virtually* some property P (e.g. cyclic, torsion-free, solvable, ...) if it possesses a finite index subgroup which has the property P.

Given a G-space X and a subgroup $H \leq G$, we denote by

$$X^H = \{x \in X \colon hx = x \text{ for all } h \in H\}$$

the set of points in X which are fixed by H. Recall that X is called *weakly contractible* if every map $S^{n-1} \to X$ extends continuously to a map $D^n \to X$ for all $n \geq 0$. A different way of saying the same thing this is that X is n-*connected* for every $n \geq -1$. In particular, for the first three values of n, this means that X is nonempty, path-connected, and simply connected, respectively.

© Springer Nature Switzerland AG 2019
H. Kammeyer, *Introduction to ℓ^2-invariants*, Lecture Notes in Mathematics 2247,
https://doi.org/10.1007/978-3-030-28297-4_4

Theorem 4.2 *Let $\mathcal{F}$ be a family of subgroups of G and let E be a G-CW complex with all stabilizer groups in $\mathcal{F}$. The following are equivalent.*

(i) *For every G-CW complex X with stabilizer groups in $\mathcal{F}$ there exists a G-equivariant map $X \to E$ which is unique up to G-homotopy.*
(ii) *For all $H \in \mathcal{F}$ the fixed point set E^H is weakly contractible.*

Proof The key observation is that every $H \in \mathcal{F}$ defines a functor from the category of G-spaces to the category of spaces by sending a G-space X to the fixed point space X^H and a G-map $f: X \to Y$ to the restriction $f^H: X^H \to Y^H$. This functor has a left-adjoint which sends a space X to the G-space $G/H \times X$ and a map f to $\mathrm{id}_{G/H} \times f$. The adjoint relation says that we have a bijection $\mathrm{map}_G(G/H \times X, Y) \cong \mathrm{map}(X, Y^H)$, natural in X and Y, from G-maps to maps. Explicitly, it is given by

$$(f: G/H \times X \to Y) \longmapsto (x \mapsto f(H, x))$$

with inverse

$$(h: X \to Y^H) \longmapsto ((gH, x) \mapsto g\, h(x)).$$

Now we prove (i) $\Rightarrow$ (ii). For all $H \in F$, assertion (i) guarantees that we have a G-map $f: G/H \to E$ and hence $f(H)$ is a point in E^H showing that E^H is not empty. For $n \geq 1$, we view a map $f: S^{n-1} \to E^H$ as a point in the space $\mathrm{map}(S^{n-1}, E^H)$ with the compact-open topology. In this topology, the adjoint relation becomes a homeomorphism $\mathrm{map}(S^{n-1}, E^H) \cong \mathrm{map}_G(G/H \times S^{n-1}, E)$ and the latter space is path-connected by assumption (i). Thus we can find a path from f to any constant map $S^{n-1} \to E^H$. Such a path defines a null-homotopy of f which is the same as an extension of f from S^{n-1} to D^n.

To prove (ii) $\Rightarrow$ (i), let a G-CW complex X with stabilizers in $\mathcal{F}$ be given. We construct a G-map $f: X \to E$ inductively over the skeleta of X. To begin with, we consider $X_0 = \coprod_{i \in I_0} G/H_i$. By (ii), the spaces E^{H_i} are not empty so we can pick points $x_i \in E^{H_i}$ for all $i \in I_0$. Requiring that $H_i \in G/H_i$ map to x_i determines a G-map $G/H_i \to E$ uniquely by equivariance. The coproduct of all these G-maps gives a G-map $f_0: X_0 \to E$. Any other G-map $f_0': X_0 \to E$ defines points $x_i' = f'(H_i) \in E^{H_i}$. Together with the x_i, these give G-maps $G/H_i \times S^0 \to E$. The adjoint maps $S^0 \to E^{H_i}$ extend to D^1 by (ii) and accordingly the original maps extend to $G/H_i \times D^1$. The coproduct of these extensions defines a G-homotopy from f_0 to f_0'.

Now assume a G-map $f_{n-1}: X_{n-1} \to E$, unique up to G-homotopy, is given. We choose a G-equivariant pushout

$$
\begin{array}{ccc}
\coprod_{i \in I_n} G/H_i \times S^{n-1} & \xrightarrow{\;q_n\;} & X_{n-1} \\
\Big\downarrow{\scriptstyle i_n} & & \Big\downarrow{\scriptstyle j_n} \\
\coprod_{i \in I_n} G/H_i \times D^n & \xrightarrow{\;Q_n\;} & X_n.
\end{array}
$$

By (ii), the restriction map

$$\text{map}_G\left(\coprod_{i\in I_n} G/H_i \times D^n, E\right) = \coprod_{i\in I_n} \text{map}\left(D^n, E^{H_i}\right) \xrightarrow{i_n^*}$$

$$\longrightarrow \coprod_{i\in I_n} \text{map}\left(S^{n-1}, E^{H_i}\right) = \text{map}_G\left(\coprod_{i\in I_n} G/H_i \times S^{n-1}, E\right)$$

is surjective. Hence, we find a lift $F_{n-1} \in \text{map}_G\left(\coprod_{i\in I_n} G/H_i \times D^n, E\right)$ of $f_{n-1} \circ q_n$. By the universal property of the pushout, we thus obtain a G-map $f_n\colon X_n \to E$, extending f_{n-1} as desired.

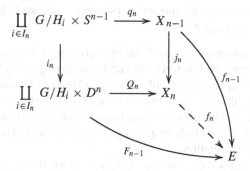

Let $f_n'\colon X_n \to E$ be another G-map. By the induction hypothesis, the restriction of f_n' to X_{n-1} is G-homotopic to f_{n-1}. Since the inclusion $X_{n-1} \to X_n$ is a cofibration, this homotopy extends to a homotopy $H\colon X_n \times I \to E$ with $H_0 = f_n'$. We may thus assume that f_n' and f_n coincide on X_{n-1}. In that case, f_n' and f_n unify to a map $G_n\colon Y_n \to E$ from the pushout Y_n of the diagram $X_n \leftarrow X_{n-1} \to X_n$. The equivariant n-cells $G/H_i \times D^n$ of X_n appear twice in Y_n as "opposite" equivariant cells so that the characteristic maps of the cells composed with G_n define G-maps $G/H_i \times S^n \to E$ where the n-disk D^n is embedded once as upper and once as lower hemisphere in S^n. Again by (ii) these maps extend to maps $G/H_i \times D^{n+1} \to E$ so that we can push the copy of X_n embedded in Y_n by upper hemispheres through $(n + 1)$-disks to the copy of X_n embedded by lower hemispheres, fixing X_{n-1} throughout. This defines a G-homotopy from f_n to f_n'. □

Definition 4.3 A G-CW complex $E_{\mathcal{F}}G$ with stabilizers in $\mathcal{F}$ satisfying conditions (i) and (ii) above is called a *classifying space of G for $\mathcal{F}$*.

The concept of classifying spaces for families of subgroups is due to T. tom Dieck.

Theorem 4.4 *For every group G and every family of subgroups $\mathcal{F}$ of G there exists a classifying space $E_{\mathcal{F}}G$ which is unique up to G-homotopy equivalence.*

Proof For the existence part of the theorem, we construct $E_{\mathcal{F}}G$ inductively over the skeleta. To begin with, set $(E_{\mathcal{F}}G)_0 = \coprod_{H \in F} G/H$. Now suppose $(E_{\mathcal{F}}G)_{n-1}$ is already given. For all $H \in \mathcal{F}$, pick one map $S^{n-1} \to ((E_{\mathcal{F}}G)_{n-1})^H$ from each homotopy class and attach an equivariant n-cell $G/H \times D^n$ according to the adjoint map $G/H \times S^{n-1} \to (E_{\mathcal{F}}G)_{n-1}$. Thus $(E_{\mathcal{F}}G)_n$ is defined for all $n \geq 0$ and we set $E_{\mathcal{F}}G = \mathrm{colim}_{n \geq 0}(E_{\mathcal{F}}G)_n$.

Note that for each $H \in \mathcal{F}$, the fixed point set $(E_{\mathcal{F}}G)^H$ forms a (non-equivariant) subcomplex as it consists of a closed union of open cells. Thus by cellular approximation, any map $S^{n-1} \to (E_{\mathcal{F}}G)^H$ is homotopic to a map $S^{n-1} \to (E_{\mathcal{F}}G^H)_{n-1} = ((E_{\mathcal{F}}G)_{n-1})^H$. By construction, this map is homotopic to the adjoint of an attaching map of an equivariant n-cell, whence it is null-homotopic.

Uniqueness follows from the usual nonsense: Say $E^1_{\mathcal{F}}G$ and $E^2_{\mathcal{F}}G$ are two classifying spaces of G for $\mathcal{F}$. By property (i), there are G-maps $E^1_{\mathcal{F}}G \to E^2_{\mathcal{F}}G$ and $E^2_{\mathcal{F}}G \to E^1_{\mathcal{F}}G$ whose compositions, by uniqueness, are G-homotopic to the identity maps on $E^1_{\mathcal{F}}G$ and $E^2_{\mathcal{F}}G$. $\qquad\square$

For the sake of a transparent proof, we have given a construction of an $E_{\mathcal{F}}G$ in the proof which cannot be functorial for family preserving group homomorphisms as we have chosen representatives of homotopy classes to obtain attaching maps. There are, however, also functorial *models* for $E_{\mathcal{F}}G$. The terminology that some G-CW complex is a model for $E_{\mathcal{F}}G$ is meant to stress that a particular G-CW complex within the uniquely defined G-homotopy equivalence class of classifying spaces of G for $\mathcal{F}$ is under consideration.

For simplicity, we set $\underline{EG} = E_{\mathcal{F}IN}G$, also called the *classifying space for proper actions*, and $EG = E_{\mathcal{T}RI\mathcal{V}}G$, called the *classifying space for free actions*. Since G acts freely on EG, the quotient map $EG \to G \backslash EG$ is a covering. The base space $G \backslash EG$ is more commonly denoted by BG and is an aspherical CW complex because by Whitehead's theorem, a weakly contractible CW complex is contractible in the usual sense. Conversely, every aspherical CW complex X is a model for $B(\pi_1 X)$. If people plainly talk about a classifying space for G, most commonly they refer to some model of BG.

Exercises

4.1.1 Let X be a connected CW complex and let Y be a model for BG. Show that every homomorphism $\pi_1(X, x_0) \to \pi_1(Y, y_0)$ is induced by a map $(X, x_0) \to (Y, y_0)$. *Hint: First assume that x_0 is the only 0-cell in X.*

4.2 Extended von Neumann Dimension

As a consequence of our discussion so far, we obtain the classical result that aspherical CW complexes are determined uniquely up to homotopy equivalence by the fundamental group. Accordingly, every homotopy invariant of CW complexes gives an isomorphism invariant of groups, such as *group homology* $H_*(G) = H_*(BG)$, *group cohomology* $H^*(G) = H^*(BG)$ and *Betti numbers of groups* $b_n(G) = \dim_{\mathbb{Q}} H_n(BG; \mathbb{Q})$. Similarly, G-homotopy invariants of G-CW complexes give invariants of groups. In a moment we will go forward and define the n-th ℓ^2-Betti number of a group G with respect to some family $\mathcal{F}$ by setting $b_n^{(2)}(G, \mathcal{F}) = b_n^{(2)}(E_{\mathcal{F}}G)$. But before we do so, we need to address one important issue. So far we defined ℓ^2-Betti numbers only for proper, finite type G-CW complexes. But given an arbitrary group G and some family $\mathcal{F}$, it cannot be assured that it possesses a finite type model for $E_{\mathcal{F}}G$. Moreover, if $\mathcal{F}$ is not contained in $\mathcal{FIN}$, then $E_{\mathcal{F}}G$ will never be proper.

Let us reflect on what made us require so far that any G-CW complex X under our consideration was proper and of finite type. These conditions were imposed to make sure that reduced ℓ^2-homology $H_n^{(2)}(X)$ as defined in 3.12 is a finitely generated Hilbert module so that its von Neumann dimension is defined as in 2.37. Properness was essential because Proposition 3.7 and Example 2.41 show that only in this case the ℓ^2-chain complex consists of Hilbert modules. "Finite type" makes sure these Hilbert modules are finitely generated. One can define von Neumann dimension for general Hilbert modules $H \subseteq \ell^2 G \otimes K$ by adding up von Neumann traces diagonally after choosing an orthonormal basis of K. This might of course give infinite values. However, given a proper, infinite type G-CW complex, it is not clear how to come up with a Hilbert module structure on ℓ^2-homology. One cannot expect anymore that the differentials $d_n^{(2)}$ in the cellular ℓ^2-chain complex are bounded operators and so it is for example not guaranteed that the ℓ^2-cycles $\ker d_n^{(2)}$ form closed subspaces.

In what follows we will explain the remedy to these technical difficulties. While we give the complete construction, proof ideas are only indicated in this section in order to not lose track on our way to the definition of ℓ^2-Betti numbers of groups. For a detailed account, we refer to [117, Sections 6.1 and 6.2].

The first insight is that once a finitely generated Hilbert module H is realized as a closed G-invariant subspaces $H \subseteq (\ell^2 G)^n$, the orthogonal projection p to this subspace is G-equivariant and thus lies in the amplified group von Neumann algebra, $p \in M_n(\mathcal{R}(G))$ by Example 2.28. This projection, in turn, gives rise to a left $\mathcal{R}(G)$-submodule $\mathcal{R}(G)^n p$ of $\mathcal{R}(G)^n$. The point is that $\mathcal{R}(G)^n p$ is a *finitely generated projective $\mathcal{R}(G)$-module* in the algebraic sense: it is complemented as a direct summand in a free, finite rank $\mathcal{R}(G)$-module:

$$\mathcal{R}(G)^n = \mathcal{R}(G)^n p \oplus \mathcal{R}(G)^n (1 - p),$$

where $1 \in M_n(\mathcal{R}(G))$ is the unit matrix. Here we consider $\mathcal{R}(G)$ as a ring with involution "$*$" only, dismissing all topology. Conversely, given a finitely generated projective left $\mathcal{R}(G)$-module P, it is by definition complemented in some $\mathcal{R}(G)^n$ which means we can find $p \in M_n(\mathcal{R}(G))$ with $p^2 = p$ and an $\mathcal{R}(G)$-isomorphism $u : P \xrightarrow{\sim} \operatorname{im} p$ to the image of p in $\mathcal{R}(G)^n$. A priori, the idempotent p might be an oblique projection as operator on $(\ell^2 G)^n$ by right multiplication. But the orthogonal projection onto the image of p is realized by a conjugate of p in $M_n(\mathcal{R}(G))$. So we can additionally assume $p = p^*$ if we want to. In any case, the image of p as operator on $(\ell^2 G)^n$ is a finitely generated left Hilbert $\mathcal{L}(G)$-module.

It turns out that these two constructions are inverses of one another up to isomorphism. But of course we want them to be inverses up to *natural* isomorphism so that the construction ought to be functorial. The latter construction can be made functorial as follows. The element $p \in M_n(\mathcal{R}(G))$ and the isomorphism u give rise to an inner product on the $\mathbb{C}$-vector space P defined by

$$\langle x, y \rangle = \sum_{i=1}^{n} \operatorname{tr}_{\mathcal{R}(G)}(u(x)_i^* u(y)_i).$$

It is easily verified that this inner product is independent of the choices of p and u. We have $\langle gx, gy \rangle = \langle x, y \rangle$ for all $g \in G$ because G acts diagonally on $\mathcal{R}(G)^n$. Let $\overline{P}$ be the Hilbert space completion of P. The reader should convince herself that $\overline{\mathcal{R}(G)} \cong \ell^2 G$ which is a special case of the so called *GNS-construction*. Setting $Q = \ker p$, we observe that

$$\overline{P} \subseteq \overline{P} \oplus \overline{Q} \cong \overline{P \oplus Q} \cong \overline{\mathcal{R}(G)^n} \cong \overline{\mathcal{R}(G)}^n \cong (\ell^2 G)^n.$$

Hence $\overline{P}$ is a finitely generated Hilbert $\mathcal{L}(G)$-module. Here it was important that we agreed in Definition 2.34 that the embedding is not part of the structure of a Hilbert module because the above embedding depends on the choice of p.

Next we have to explain what our functor does with morphisms. Let $f : P_1 \to P_2$ be a homomorphism of finitely generated projective left $\mathcal{R}(G)$-modules. We argue that f extends continuously to a G-equivariant operator $\overline{f} : \overline{P_1} \to \overline{P_2}$. After choosing complements $P_1 \oplus Q_1 = \mathcal{R}(G)^n$ and $P_2 \oplus Q_2 = \mathcal{R}(G)^n$ for large enough n, we can extend f trivially on Q_1 and obtain an endomorphism $F : \mathcal{R}(G)^n \to \mathcal{R}(G)^n$ of a free $\mathcal{R}(G)$-module. Hence it is given by right multiplication with a matrix in $M_n(\mathcal{R}(G))$. The matrix also describes the extension $\overline{F} : (\ell^2 G)^n \to (\ell^2 G)^n$ of F to the Hilbert completions. Since the amplified group von Neumann algebra $M_n(\mathcal{R}(G))$ acts by bounded G-equivariant operators, so does the restriction of $\overline{F}$ to $\overline{P_1} \subseteq (\ell^2 G)^n$ which agrees with f on the dense subspace P_1. Hence this restriction gives the unique continuous extension $\overline{f} : \overline{P_1} \to \overline{P_2}$ of f as desired.

Theorem 4.5 *Completion defines an equivalence from the category of finitely generated, projective left $\mathcal{R}(G)$-modules to the category of finitely generated Hilbert $\mathcal{L}(G)$-modules.*

We leave the proof to the reader as the guided Exercise 4.2.1. The theorem can be made a little more precise by saying that completion is an equivalence of $\mathbb{C}$-*categories*. This essentially says that morphism sets form complex vector spaces and the equivalence preserves this structure. The theorem tells us that the category of Hilbert modules can be fully embedded into the category of left $\mathcal{R}(G)$-modules as the subcategory of finitely generated projective left $\mathcal{R}(G)$-modules. Thus the following extension of von Neumann dimension to completely general left $\mathcal{R}(G)$-modules arises naturally.

Definition 4.6 The *(extended) von Neumann dimension* of a left $\mathcal{R}(G)$-module N is given by

$$\dim_{\mathcal{R}(G)} N = \sup \{\dim_{\mathcal{R}(G)} \overline{P}: \ P \subseteq N \text{ finitely generated, projective}\}.$$

In addition to the standard properties normalization and additivity, von Neumann dimension of $\mathcal{R}(G)$-modules comes with two regularity properties. The first property goes by the name of *cofinality*, and says that von Neumann dimension equals the least upper bound of the dimensions of any exhausting family of submodules. The second, *continuity*, says that the dimension of a submodule N of some finitely generated $\mathcal{R}(G)$-module M agrees with the dimension of the *closure* $\widehat{N}$ of N in M:

$$\widehat{N} = \{x \in M: \varphi(x) = 0 \text{ for all } \varphi \in \operatorname{Hom}_{\mathcal{R}(G)}(M, \mathcal{R}(G)) \text{ with } N \subseteq \ker \varphi\}.$$

To be completely precise, we collect these properties in a theorem.

Theorem 4.7 (Properties of Extended von Neumann Dimension)

(i) Normalization. *We have* $\dim_{\mathcal{R}(G)}(\mathcal{R}(G)) = 1$.
(ii) Additivity. *Suppose* $0 \to L \to M \to N \to 0$ *is a short exact sequence of left* $\mathcal{R}(G)$-*modules. Then*

$$\dim_{\mathcal{R}(G)} M = \dim_{\mathcal{R}(G)} L + \dim_{\mathcal{R}(G)} N.$$

Here it is understood that $x + y = \infty$ *if* $x = \infty$ *and/or* $y = \infty$.
(iii) Cofinality. *Let M be a left $\mathcal{R}(G)$-module and suppose $M \cong \operatorname{colim}_i M_i$ for a system $(M_i)_{i \in I}$ of submodules of M directed by inclusion. Then*

$$\dim_{\mathcal{R}(G)} M = \sup_{i \in I} \dim_{\mathcal{R}(G)} M_i.$$

(iv) Continuity. *Let M be a finitely generated left $\mathcal{R}(G)$-module and let $N \subseteq M$ be a submodule. Then*

$$\dim_{\mathcal{R}(G)} N = \dim_{\mathcal{R}(G)} \widehat{N}$$

The proof is routine. Note that faithfulness fails for the extended von Neumann dimension. But again, this is not a bug. It's a feature that makes von Neumann dimension strikingly reminiscent to the $\mathbb{Z}$-rank of a finitely generated abelian group.

For a generic finitely generated $\mathcal{R}(G)$-module M we have a decomposition $M \cong$ $\mathbf{P}M \oplus \mathbf{T}M$, where for the *torsion part* $\mathbf{T}M = \widehat{\{0\}}$, we have $\dim_{\mathcal{R}(G)}(\mathbf{T}M) = 0$ by the continuity property (iv) and for the *projective part* $\mathbf{P}(M) = M/\mathbf{T}M$, we have $\dim_{\mathcal{R}(G)} \mathbf{P}M = \dim_{\mathcal{R}(G)} M$ by the additivity property (ii). In view of these observations, Lück goes as far as to say that the ring $\mathcal{R}(G)$ is "very similar" to $\mathbb{Z}$. Except that typically, it is not commutative, not Noetherian, and has zero divisors. In any case it is undeniable that to a large extent the module categories over $\mathbb{Z}$ and $\mathcal{R}(G)$ have a parallel structure theory. One could argue that the term *von Neumann rank* instead of "extended von Neumann dimension" would be more consistent with the above observations. But the terminology "extended von Neumann dimension" has become standard in the literature. What makes the category of $\mathcal{R}(G)$-modules well-behaved from a technical point of view, is that the ring $\mathcal{R}(G)$ is *semihereditary* which means that the property of being a projective (left) module is robust in the following three senses.

Proposition 4.8 (Semiheredity of the Ring $\mathcal{R}(G)$)

 (i) Projective submodules. *Every finitely generated submodule U of a projective $\mathcal{R}(G)$-module P is projective.*
 (ii) Projective quotients. *If V is a finitely generated $\mathcal{R}(G)$-module and $U \subseteq V$ is a submodule, then V/U is finitely generated projective.*
 (iii) Projective kernels. *If $f : Q \to P$ is a morphism of finitely generated projective $\mathcal{R}(G)$-modules, then $\ker f$ is finitely generated projective.*

Note that setting $U = 0$ in the second part of the Proposition gives the decomposition of V into projective and torsion part.

Proof We only show that (ii) implies (iii). If in (ii) the module V is also projective, then not only V/U but also $\widehat{U}$ itself is finitely generated projective because it is a direct summand in V and hence a direct summand in a finite rank free module. It is thus enough to show that $\widehat{\ker f} = \ker f$. Since P is a submodule of some $\mathcal{R}(G)^n$, the projections p_i onto the n coordinates define n linear functionals on P whose kernels have trivial intersection. Thus if $x \in \widehat{\ker f}$, then $(p_i \circ f)(x) = 0$ for $i = 1, \dots, n$ which implies $f(x) = 0$. $\square$

Exercises

4.2.1 Prove Theorem 4.5 along the following lines.

 (i) Show that the full subcategories of finite rank, free left $\mathcal{R}(G)$-modules and finite rank, free Hilbert $\mathcal{L}(G)$-modules are equivalent.
 (ii) Identify the category of finitely generated projective modules in each case with the *Karoubi envelope* (check the literature!) of these subcategories. This yields an equivalence of the categories of finitely generated projective Hilbert and $\mathcal{R}(G)$-modules.
(iii) Show that completion is naturally isomorphic to this equivalence.

4.2.2 A sequence $U \xrightarrow{f} V \xrightarrow{g} W$ of finitely generated projective left $\mathcal{R}(G)$-modules is called *weakly exact* at V if im $f \subset \ker g$ and if for any other sequence $Q \xrightarrow{v} V \xrightarrow{u} W$ of finitely generated projective $\mathcal{R}(G)$-modules we have im $v \subseteq \ker u$ whenever im $f \subseteq \ker u$ and im $v \subseteq \ker g$. Show that the equivalence of Theorem 4.5 preserves exact and weakly exact sequences.

4.3 ℓ^2-Betti Numbers of G-Spaces

Now that $\dim_{\mathcal{R}(G)}$ is defined for general left $\mathcal{R}(G)$-modules, we can define ℓ^2-Betti numbers for general G-spaces with no constraints whatsoever on the type of the space or the occurring isotropy groups. The left G-action on X induces a left $\mathbb{Z}G$-module structure on the singular chain complex $C_*^{\mathrm{sing}}(X)$. The $\mathcal{R}(G)$-$\mathbb{Z}G$-bimodule $\mathcal{R}(G)$ determines the functor $\mathcal{R}(G) \otimes_{\mathbb{Z}G}$ which turns $C_*^{\mathrm{sing}}(X)$ into a chain complex of left $\mathcal{R}(G)$-modules. So we can consider the extended von Neumann dimension of the homology.

Definition 4.9 Let X be a G-space. The *n-th ℓ^2-Betti number* of X is

$$b_n^{(2)}(X) = b_n^{(2)}(G \curvearrowright X) = \dim_{\mathcal{R}(G)} H_n(\mathcal{R}(G) \otimes_{\mathbb{Z}G} C_*^{\mathrm{sing}}(X)) \in [0, \infty].$$

It remains to check this notion is consistent with the previous definition.

Theorem 4.10 *Let X be a proper, finite type G-CW complex. Then the ℓ^2-Betti numbers of X according to Definition 3.13 coincide with the ℓ^2-Betti numbers of X according to Definition 4.9.*

Proof By [116, Lemma 4.2] one can construct a $\mathbb{Z}G$-chain homotopy equivalence $C_*^{\mathrm{cell}}(X) \to C_*^{\mathrm{sing}}(X)$ which induces an $\mathcal{R}(G)$-isomorphism

$$H_n(\mathcal{R}(G) \otimes_{\mathbb{Z}G} C_*^{\mathrm{cell}}(X)) \cong H_n(\mathcal{R}(G) \otimes_{\mathbb{Z}G} C_*^{\mathrm{sing}}(X)).$$

Since

$$\dim_{\mathcal{R}(G)} H_n(\mathcal{R}(G) \otimes_{\mathbb{Z}G} C_*^{\mathrm{cell}}(X)) = \dim_{\mathcal{R}(G)} \mathbf{P}H_n(\mathcal{R}(G) \otimes_{\mathbb{Z}G} C_*^{\mathrm{cell}}(X)),$$

it remains to show that

$$\overline{\mathbf{P}H_n(\mathcal{R}(G) \otimes_{\mathbb{Z}G} C_*^{\mathrm{cell}}(X))} \cong H_n^{(2)}(X).$$

To this end, consider the differentials $1 \otimes d_*$ of the chain complex $\mathcal{R}(G) \otimes_{\mathbb{Z}G} C_*^{\mathrm{cell}}(X)$. As functionals $H_n(\mathcal{R}(G) \otimes_{\mathbb{Z}G} C_*^{\mathrm{cell}}(X)) \to \mathcal{R}(G)$ are the same as functionals on $\ker 1 \otimes d_n$ vanishing on im $1 \otimes d_{n+1}$, the sequence

$$0 \to \widehat{\mathrm{im}\, 1 \otimes d_{n+1}} \to \ker 1 \otimes d_n \to \mathbf{P}H_n(\mathcal{R}(G) \otimes_{\mathbb{Z}G} C_*^{\mathrm{cell}}(X)) \to 0$$

is exact. The point is that by Proposition 4.8 and its proof, all three $\mathcal{R}(G)$-modules in this sequence are finitely generated projective, so that our completion functor $\overline{(\,\cdot\,)}$ is available. This functor is an equivalence so it preserves exact sequences. From this we obtain a diagram with exact rows

$$
\begin{array}{ccccccccc}
0 & \longrightarrow & \overline{\mathrm{im}\,1 \otimes d_{n+1}} & \longrightarrow & \overline{\ker 1 \otimes d_n} & \longrightarrow & \overline{PH_n(\mathcal{R}(G) \otimes_{\mathbb{Z}G} C_*^{\mathrm{cell}}(X))} & \longrightarrow & 0 \\
 & & \downarrow{\cong} & & \downarrow{\cong} & & \downarrow & & \\
0 & \longrightarrow & \overline{\mathrm{im}\,d_{n+1}^{(2)}} & \longrightarrow & \ker d_n^{(2)} & \longrightarrow & H_n^{(2)}(X) & \longrightarrow & 0
\end{array}
$$

in which the first two vertical arrows define the third. The bars in the first line denote the completion functor, whereas the bar in the lower left means closure of a subspace in Hilbert space. Since the first two vertical arrows are isomorphisms, so is the third by the five lemma.   □

Theorem 4.11 (Computation of ℓ^2-Betti Numbers Revisited)

   (i) Homotopy invariance. Let $f : X \to Y$ be a G-homotopy equivalence of G-spaces X and Y. Then $b_n^{(2)}(X) = b_n^{(2)}(Y)$ for all $n \geq 0$.

   (ii) Zeroth ℓ^2-Betti number. Let X be a path-connected G-space. Then $b_0^{(2)}(X) = \frac{1}{|G|}$ with $\frac{1}{\infty} = 0$.

   (iii) Künneth formula. Let X_1 and X_2 be G_1- and G_2-spaces, respectively. Then $X_1 \times X_2$ is a $G_1 \times G_2$-space and for all $n \geq 0$ we have

$$
b_n^{(2)}(X_1 \times X_2) = \sum_{p+q=n} b_p^{(2)}(X_1)\, b_q^{(2)}(X_2).
$$

   (iv) Restriction. Let X be a G-space and let $G_0 \leq G$ be a finite index subgroup. Then $\mathrm{res}_{G_0}^G X$ is a G_0-space and $b_n^{(2)}(\mathrm{res}_{G_0}^G X) = [G : G_0] b_n^{(2)}(X)$ for all $n \geq 0$.

Comparing this theorem to the previous cellular version Theorem 3.18, you will notice that it is verbatim the same result except that "proper, finite type G-CW complex" now simply reads "G-space" and "connected" was replaced by "path-connected". The latter two notions might differ for general spaces whereas for CW complexes they do not. In (iii), the rules $x+\infty = \infty$, $x \cdot \infty = \infty$ and $\infty \cdot 0 = 0$ apply if some occurring ℓ^2-Betti number is infinite. We content ourselves with having proven the cellular version and skip the proof of this generalization which the reader may find in [116] and [117, Theorem 6.54].

4.4 ℓ^2-Betti Numbers of Groups and How to Compute Them

Now that we have defined ℓ^2-Betti numbers of general G-spaces, we are in the position to give the following definition.

Definition 4.12 Let G be a group and let $\mathcal{F}$ be a family of subgroups. For $n \geq 0$, the *n-th ℓ^2-Betti number* of G with respect to $\mathcal{F}$ is defined by

$$b_n^{(2)}(G, \mathcal{F}) = b_n^{(2)}(E_{\mathcal{F}} G) \in [0, \infty].$$

Let us moreover agree that $b_n^{(2)}(G)$ shall mean $b_n^{(2)}(G, \mathcal{TRIV}) = b_n^{(2)}(EG)$.

Example 4.13 One can see geometrically that every finite group G has a finite type model for the classifying space EG. Since EG is contractible, it follows from Theorem 4.10 and Example 3.14 that $b_n^{(2)}(G) = 0$ for all $n \geq 1$ and $b_0^{(2)}(G) = 1/|G|$. Note that you will prove in Exercise 4.4.1 that a finite group G has no finite model for EG unless G is trivial. In Exercises 4.4.2 and 4.4.3 you will show more precisely that in fact EG has not even a finite-dimensional model.

As opposed to G-spaces, it is convenient to say that a group G is ℓ^2-*acyclic* if $b_n^{(2)}(G) = 0$ for $n \geq 1$ so that this notion includes the finite groups. We can now clarify the role of the family $\mathcal{F}$.

Theorem 4.14 *Let G be a group and let $\mathcal{F}$ be a family consisting of ℓ^2-acyclic subgroups. Then $b_n^{(2)}(G, \mathcal{F}) = b_n^{(2)}(G)$ for all $n \geq 0$.*

Proof We only give an outline. By Theorem 4.2 we have a G-equivariant map $EG \rightarrow E_{\mathcal{F}} G$ which is unique up to G-homotopy. We can turn this into an equivariant map of free G-spaces by replacing $E_{\mathcal{F}} G$ with $E_{\mathcal{F}} G \times EG$ on which G acts diagonally ("the *Borel construction*") and consider instead the diagonal map $EG \rightarrow E_{\mathcal{F}} G \times EG$. One can show that applying $\mathcal{R}(G) \otimes_{\mathbb{Z}G} C_*^{\mathrm{sing}}(\cdot)$ gives a chain map that induces an isomorphism in homology and that the Borel construction does not alter ℓ^2-Betti numbers.   □

Theorem 4.14 says that instead of the trivial family, we can alternatively use the families $\mathcal{FIN}$, $\mathcal{VCYC}$ and—as information for the initiated reader, see Sect. 5.4.6—even the family $\mathcal{AME}$ of *amenable* subgroups to compute the ℓ^2-Betti numbers of G. However, in practically all cases of interest it turns out that $\mathcal{FIN}$ is the best choice. Firstly, EG often has finite models when the others have not, and secondly, if EG is at least finite type, the cellular definition (Definition 3.13) of ℓ^2-Betti numbers still applies because EG is by definition proper. For example, the real line turned into a CW complex with 0-cells at the half-integers allows a cellular action by the infinite dihedral group $D_\infty = \mathbb{Z} \rtimes \mathbb{Z}/2\mathbb{Z}$. The reader may convince herself that this defines a model for ED_∞. Recall that by Exercise 4.4.3, no group with torsion elements can have a finite-dimensional model for EG.

Table 4.1 Examples of classifying spaces and ℓ^2-Betti numbers

G	BG	EG	$\underline{E}G$	$b_n(G)$	$b_n^{(2)}(G)$
$\mathbb{Z}$	S^1	$\mathbb{R}$	$\mathbb{R}$	$\begin{cases} 1 & n=0,1 \\ 0 & n\geq 2 \end{cases}$	$0 \quad n\geq 0$
$\mathbb{Z}^k$	$\mathbb{T}^k$	$\mathbb{R}^k$	$\mathbb{R}^k$	$\binom{k}{n}$	$0 \quad n\geq 0$
F_k	$\bigvee_{i=1}^k S^1$	k-reg. tree	k-reg. tree	$\begin{cases} 1 & n=0 \\ k & n=1 \\ 0 & n\geq 2 \end{cases}$	$\begin{cases} 0 & n\neq 1 \\ k-1 & n=1 \end{cases}$
F_∞	$\bigvee_{i=1}^\infty S^1$	∞-reg. tree	∞-reg. tree	$\begin{cases} 1 & n=0 \\ \infty & n=1 \\ 0 & n\geq 2 \end{cases}$	$\begin{cases} 0 & n\neq 1 \\ \infty & n=1 \end{cases}$
$\mathbb{Z}/2$	$\mathbb{R}P^\infty$	S^∞	$\bullet$	$\begin{cases} 1 & n=0 \\ 0 & n\geq 1 \end{cases}$	$\begin{cases} 1/2 & n=0 \\ 0 & n\geq 1 \end{cases}$
D_∞	$\mathbb{R}P^\infty \vee \mathbb{R}P^\infty$	See text	$\mathbb{R}$	$\begin{cases} 1 & n=0 \\ 0 & n\geq 1 \end{cases}$	$0 \quad n\geq 0$
$PSL(2,\mathbb{Z})$	$\mathbb{R}P^\infty \vee \mathbb{Z}/3\backslash S^\infty$	See text	$\mathbb{H}^2$	$\begin{cases} 1 & n=0 \\ 0 & n\geq 1 \end{cases}$	$\begin{cases} 1/6 & n=1 \\ 0 & n\neq 1 \end{cases}$

Table 4.1 gives examples of Betti numbers and ℓ^2-Betti numbers of various groups. The reader is advised to check it entry for entry. In the second to last example, one can picture the space ED_∞, which is the universal covering of $\mathbb{R}P^\infty \vee \mathbb{R}P^\infty$, as follows. The base point in $\mathbb{R}P^\infty$ has two lifts in S^∞, call them left and right. We line up countably many copies of S^∞ and identify the right base point of each copy with the left base point of the succeeding copy. A generator of the infinite cyclic normal subgroup $\mathbb{Z} \trianglelefteq D_\infty = \mathbb{Z} \rtimes \mathbb{Z}/2\mathbb{Z}$ acts on this space by translating the copies by two steps. The subgroup $\mathbb{Z}/2\mathbb{Z} \leq D_\infty$ acts by inversion about the center of a certain copy of S^∞ so that on this particular copy it acts as the antipodal map and maps the n-th neighbor on the right homeomorphically to the n-th neighbor on the left.

In the last example, a matrix $\pm \begin{pmatrix} a & b \\ c & d \end{pmatrix} \in PSL(2,\mathbb{Z})$ acts on the upper half plane model of $\mathbb{H}^2$ by $z \mapsto (az+b)/(cz+d)$. Note that the action is not cocompact so that this model of $\underline{E}G$ is not of finite type. The value $b_1^{(2)}(PSL(2,\mathbb{Z})) = 1/6$ can be obtained by analytic methods (see the remarks below Theorem 3.39) and a careful comparison of analytic and cellular ℓ^2-Betti numbers. Way more easily, however, one can compute this value by means of the well-known isomorphism $PSL(2,\mathbb{Z}) \cong \mathbb{Z}/2 * \mathbb{Z}/3$ and Theorem 4.15 (ii) below. Due to this isomorphism, a model for $BPSL(2,\mathbb{Z})$ is given by $\mathbb{R}P^\infty \vee (\mathbb{Z}/3\mathbb{Z} \backslash S^\infty)$, where you will construct the action of $\mathbb{Z}/3\mathbb{Z}$ on S^∞ in Exercise 4.4.2. The universal covering $EPSL(2,\mathbb{Z})$ can hence again be obtained by gluing countably many copies of S^∞. This time, however, these are glued so as to be aligned along a 3-regular tree, where the copies of S^∞ covering $\mathbb{RP}^\infty$ form the edges and the copies of S^∞ covering $\mathbb{Z}/3\mathbb{Z} \setminus S^\infty$ form the vertices of the tree.

Theorem 4.15 *The following formulas hold for any groups.*

(i) $b_n^{(2)}(G_1 \times G_2) = \sum_{p+q=n} b_p^{(2)}(G_1) \cdot b_q^{(2)}(G_2)$ *for all* $n \geq 0$.
(ii) For the free product of two groups, we have

$$b_n^{(2)}(G_1 * G_2) = b_n^{(2)}(G_1) + b_n^{(2)}(G_2) \text{ for } n \geq 2,$$

$$b_1^{(2)}(G_1 * G_2) = 1 + b_1^{(2)}(G_1) - \frac{1}{|G_1|} + b_1^{(2)}(G_2) - \frac{1}{|G_2|},$$

$$b_0^{(2)}(G_1 * G_2) = 0 \text{ if } G_1 \text{ and } G_2 \text{ are nontrivial.}$$

(iii) $b_n^{(2)}(G_0) = [G : G_0] \, b_n^{(2)}(G)$ *for all* $n \geq 0$ *if* $G_0 \leq G$ *has finite index.*

Proof The product $G_1 \times G_2$ acts freely on the product space $EG_1 \times EG_2$. A map $S^{n-1} \to EG_1 \times EG_2$ is the same as two maps $S^{n-1} \to EG_i$ for $i = 1, 2$. These extend to D^n, hence $E(G_1 \times G_2) \simeq EG_1 \times EG_2$. So (i) follows from the Künneth formula for ℓ^2-Betti numbers of G-spaces, Theorem 4.11 (iii). For (ii), note that $E(G_1 * G_2) \simeq \widetilde{BG_1 \vee BG_2}$ by van Kampen's theorem and observing that $BG_1 \vee BG_2$ arises by gluing alternately the weakly contractible spaces EG_1 and EG_2 in a tree-like pattern. The asserted formulas follow from a Mayer–Vietoris type argument which we will skip as it is technically somewhat involved. Since we have $EG_0 \simeq \text{res}_{G_0}^G EG$, we see that (iii) follows from Theorem 4.11 (iv).   □

Recall from Table 4.1 that $b_1^{(2)}(\text{PSL}(2, \mathbb{Z})) = 1/6$ and $b_1^{(2)}(F_2) = 1$. Thus part (iii) of the above theorem has the curious consequence that any embedding $F_2 \subset \text{PSL}(2, \mathbb{Z})$ either has infinite index or index six. This can also be seen by an Euler characteristic argument.

The groups G we considered so far either possess a finite type model for EG or are not even finitely generated. Let us conclude this section by pointing the reader to a result of Lück and Osin [121, Theorem 4.1] who construct for every $n \geq 2$ and every $\varepsilon > 0$ an infinite p-group Q with n generators such that $b_1^{(2)}(Q) \geq n - 1 - \varepsilon$. The group Q is not finitely presented and hence does not admit a model for EQ with finite 2-skeleton. It arises as a colimit of certain quotients of the free group F_n which lie in a *pro-p group* and whose first ℓ^2-Betti number can be uniformly bounded from below by $n - 1 - \varepsilon$. The bound is still valid for Q by M. Pichot's semicontinuity theorem for the first ℓ^2-Betti number [144, Theorem 1.1]. The group Q is of interest because it shows that the finite type assumption in Lück's approximation theorem, to be examined in Chap. 5, cannot be omitted as we will discuss in Sect. 5.4.1.

Exercises

4.4.1 Let G be a finite group. Show that G has no finite model for BG unless G is trivial. *Hint: Euler characteristic.*

4.4.2 Show that $S^\infty = \mathrm{colim}_n S^n$ is contractible by constructing an explicit homotopy from the identity map of S^∞ to a constant map. For each $m \geq 2$ find a CW structure on S^∞ and a free, cellular action of $\mathbb{Z}/m\mathbb{Z}$ on S^∞ to obtain $(\mathbb{Z}/m\mathbb{Z})\backslash S^\infty$ as a model for $B(\mathbb{Z}/m\mathbb{Z})$. *Hint: Review the homology computation of lens spaces.*

4.4.3 Show that for all $n \geq 1$ we have $b_n(\mathbb{Z}/m\mathbb{Z}) = 0$ whereas $H_n(\mathbb{Z}/m\mathbb{Z}; \mathbb{Z}/m\mathbb{Z}) \cong \mathbb{Z}/m\mathbb{Z}$. Conclude that a group G which has a finite-dimensional model for BG is torsion-free. In particular, a finite group G has no finite-dimensional model for BG unless G is trivial.

4.4.4 Let $p \geq 0$ and $q, n \geq 1$ be integers.

(i) Find a group $G = G(n; p, q)$ with finite type model for EG such that $b_k^{(2)}(G) = 0$ for $k \neq n$ and $b_n^{(2)}(G) = \frac{p}{q}$.
(ii) Find a group as above that additionally satisfies $b_k(G) = 0$ for $k \geq 1$.

4.5 Applications of ℓ^2-Betti Numbers to Group Theory

With the concept of ℓ^2-Betti numbers of groups at hand, we will present some interesting applications and relations to other concepts of group theory. Once again, ℓ^2-Betti numbers will have things to say in contexts that are not related to ℓ^2-methods in any apparent way.

4.5.1 Detecting Finitely Co-Hopfian Groups

Definition 4.16 A group G is called (finitely) co-Hopfian if G is not isomorphic to a proper subgroup of G (of finite index).

Obviously, finite groups are co-Hopfian and free abelian groups are not.

Theorem 4.17 *Let G be a group and assume $0 < b_n^{(2)}(G) < \infty$ for some $n \geq 0$. Then G is finitely co-Hopfian.*

Proof If G has a subgroup $H \leq G$ of index $1 < [G : H] < \infty$ such that $H \cong G$, then by Theorem 4.15 (iii) we have for all $n \geq 0$ that

$$b_n^{(2)}(G) = [G : H] \, b_n^{(2)}(H) = [G : H] \, b_n^{(2)}(G)$$

and hence either $b_n^{(2)}(G) = 0$ or $b_n^{(2)}(G) = \infty$.   □

Thus nonabelian free groups F_k for $k \geq 2$ are finitely co-Hopfian and so is $PSL(2, \mathbb{Z})$. Free groups do however contain free groups of larger rank as proper finite index subgroups. To see this, consider the quotient map

which identifies pairs of points on the three circles mapped to one another by a point reflection through the center of the middle circle. It is a twofold covering map illustrating that F_2 contains F_3 as a subgroup of index 2. Say the left hand circle of the base space is the image of the two outer circles and the right hand circle is the $\mathbb{RP}^1$ image of the middle circle. Let us pick the left hand wedge point as base point of the covering space. If a and b are the two generators of the free fundamental group of the base space that wind around the left and right circle, respectively, then we see that the characteristic subgroup $F_3 \leq F_2 = \langle a, b \rangle$ is generated by a, b^2, and bab^{-1}.

Note that on the other hand F_k is not co-Hopfian. It can be embedded into itself as an infinite index subgroup in numerous ways, for example using that the commutator subgroup of F_k is isomorphic to F_∞. Of course Theorem 4.17 only gives a sufficient condition for a group to be finitely co-Hopfian. For instance, fundamental groups of closed (and more generally of finite volume) hyperbolic 3-manifolds are ℓ^2-acyclic, as we discussed in Theorem 3.39, but they are also finitely co-Hopfian. One way to see this, is that these manifolds have non-zero ℓ^2-*torsion*, another multiplicative invariant which we will introduce in Chap. 6.

4.5.2   Bounding the Deficiency of Finitely Presented Groups

For a finite presentation $P = \langle S|R \rangle$ of a finitely presented group G, let $g(P) = |S|$ be the number of generators and let $r(P) = |R|$ be the number of relators in P.

Definition 4.18 The *deficiency* of G is $\mathrm{def}(G) = \max\limits_{P}\{g(P) - r(P)\}$.

Here the maximum is taken over all finite presentations of G. Intuitively, adding another generator to some presentation of G should cost another relation. So it seems plausible that the maximum above exists. To see this rigorously, we make use of the *presentation complex* X_P associated with P. This is a two-dimensional CW-complex, whose 1-skeleton is a wedge sum of as many circles S^1 as P has generators. To these we attach one 2-cell for each relation r in P by the attaching map the word r describes when we orient and label the circles by the generators s of P. By construction we have $\pi_1(X_P) \cong G$. We can kill the higher homotopy groups of X_P by attaching cells of dimension three and higher. This extends X_P to a model of BG with finite 2-skeleton. Hence the inclusion map $X_P \to BG$ induces isomorphisms $H_i(X_P) \xrightarrow{\sim} H_i(BG)$ for $i = 0, 1$ and an epimorphism $H_2(X_P) \to H_2(BG)$. It follows that

$$g(P) - r(P) = 1 - \chi(X_P) = 1 - b_0(X_P) + b_1(X_P) - b_2(X_P)$$
$$= b_1(X_P) - b_2(X_P) \le b_1(G) - b_2(G).$$

Example 4.19 For the free abelian groups we have

$$\mathrm{def}(\mathbb{Z}^n) = \frac{n(3-n)}{2} = n - \binom{n}{2}.$$

Indeed, the presentation $\mathbb{Z}^n = \langle x_1, \ldots, x_n \mid [x_i, x_j] \, 1 \le i < j \le n \rangle$ gives the inequality "$\ge$" and the reverse inequality "$\le$" follows from the calculation above. Similarly we see $\mathrm{def}(F_n) = n$.

One can do the exact same calculation as above for ℓ^2-homology to conclude the following result.

Theorem 4.20 *Let G be any finitely presented group. Then*

$$\mathrm{def}(G) \le 1 - b_0^{(2)}(G) + b_1^{(2)}(G) - b_2^{(2)}(G).$$

Depending on G, this bound can be better or worse than the one given by ordinary Betti numbers. For example for $G = \mathrm{PSL}(2, \mathbb{Z})$, Theorem 4.20 only gives $\mathrm{def}(\mathrm{PSL}(2, \mathbb{Z})) \le 7/6$, hence $\mathrm{def}(\mathrm{PSL}(2, \mathbb{Z})) \le 1$ whereas the ordinary Betti number bound is $\mathrm{def}(\mathrm{PSL}(2, \mathbb{Z})) \le 0$. This is the correct value because $\mathrm{PSL}(2, \mathbb{Z}) = \langle a, b \mid a^2, b^3 \rangle$. On the other hand, let $G \le \mathrm{Isom}(\mathbb{H}^4)$ be a discrete, torsion-free subgroup such that the quotient space $G \backslash \mathbb{H}^4$ is a hyperbolic 4-manifold of finite volume. In that case one can find a CW structure on $\mathbb{H}^4$ such that the free action of G is cellular and cocompact. Since $\mathbb{H}^4$ is contractible, this gives a finite model for EG. Recall that we mentioned on p. 64 that a hyperbolic 4-manifold only has one non-zero ℓ^2-Betti number which sits in degree 2. Hence Theorems 4.20 and 3.19 give $\mathrm{def}(G) \le 1 - \chi(G)$. *Hirzebruch's proportionality principle* [76] says that $\chi(G)$ is proportional to the volume of $G \backslash \mathbb{H}^4$ and the proportionality constant in this case is given by the ratio of Euler characteristic and volume of the 4-sphere S^4. So we get $\mathrm{def}(G) \le 1 - \frac{3}{4\pi^2} \mathrm{vol}(G \backslash \mathbb{H}^4)$ as observed by Lott [112]. Since G has subgroups of arbitrarily large index, as we will discuss in Sect. 5.1 of Chap. 5, it follows that G has subgroups with arbitrarily large negative deficiency.

4.5.3 One Relator Groups and the Atiyah Conjecture

Let G be a finitely generated *one relator group*, meaning a group with presentation $P = \langle x_1, \ldots x_g \mid r \rangle$ in which the nontrivial word r in the letters $x_1, \ldots, x_g$ is the only relator. If G is torsion-free, so that of necessity $g \geq 2$, it is known that the presentation complex X_P is already aspherical and hence a two-dimensional model for BG [124, III, Paragraph 9-11]. It follows that $\mathrm{def}(G) = g - 1$ because

$$g - 1 \leq \mathrm{def}(G) \leq b_1(G) - b_2(G) = -\chi(G) + 1 = -(1 - g + 1) + 1 = g - 1.$$

For the ℓ^2-homology of G, Lück [117, p. 301] expected (compare the remarks by Gromov in [64, 8.A4]) and Dicks–Linnell [32] confirmed the following.

Theorem 4.21 *Let G be a torsion-free group with g generators and one relator. Then $b_1^{(2)}(G) = g - 2$ and $b_2^{(2)}(G) = 0$.*

We saw in Example 3.26 that the theorem is true for surface groups. The proof of Theorem 4.21 in [32] uses Linnell's theorem 3.33 on the Atiyah conjecture. We will however report at the end of Chap. 5 that very recently, the Atiyah conjecture was proven for torsion-free one relator groups which allows an easier conclusion of the theorem, as was observed by Lück.

Proof Since the classifying space $BG = X_P$ has no 3-cells, we have

$$H_2^{(2)}(G) = \ker\left(d_2^{(2)} \colon C_2^{(2)}(\widetilde{X}_P) \longrightarrow C_1^{(2)}(\widetilde{X}_P)\right)$$

and moreover $C_2^{(2)}(\widetilde{X}_P) \cong \ell^2 G$ because X_P has precisely one 2-cell corresponding to the only relator in P. Since the (reduced) relator word r is nontrivial by definition, it follows that the homomorphism $d_2^{(2)}$ is nontrivial. Hence $\ker d_2^{(2)}$ is a proper Hilbert submodule of $C_2^{(2)}(\widetilde{X}_P)$ which implies $0 \leq b_2^{(2)}(G) = \dim_{\mathcal{R}(G)} \ker d_2^{(2)} < 1$. As the Atiyah conjecture 3.30 is true for G with coefficients in $\mathbb{Z}$, it follows that $b_2^{(2)}(G) = 0$. Theorem 3.19 gives

$$b_1^{(2)}(G) = -\chi(X_P) + b_0^{(2)}(G) + b_2^{(2)}(G) = -(1 - g + 1) = g - 2. \qquad \square$$

4.5.4 The Zeroth ℓ^2-Betti Number

We can finally settle a debt and give the missing part of the proof of Theorem 3.18 (ii), namely that $b_0^{(2)}(X) = 0$ if G is an infinite group and X is any connected, nonempty, proper, finite type G-CW complex X. By Theorems 4.2 and 4.4 we have a G-map $f \colon X \to \underline{EG}$, unique up to G-homotopy. Similar to the proof of Theorem 4.14, we can go over to the Borel construction and consider the G-map id $\times f: EG \times X \to EG \times \underline{EG}$ of free G-CW complexes. Since X is connected and nonempty, the induced map $H_0(\text{id} \times f; \mathbb{C})$ in (singular) homology is an isomorphism and $H_1(\text{id} \times f; \mathbb{C})$ is trivially surjective because $H_1(EG \times \underline{EG}; \mathbb{C}) = 0$. In other words id $\times f$ is *homologically 1-connected* and a homological algebra argument shows that it remains homologically 1-connected if coefficients are taken in $\mathcal{R}(G)$ instead of $\mathbb{C}$ [116, Lemma 4.8]. It follows that $b_0^{(2)}(EG \times X) = b_0^{(2)}(EG \times \underline{EG})$. As the Borel construction does not alter ℓ^2-Betti numbers, we obtain $b_0^{(2)}(X) = b_0^{(2)}(\underline{EG})$ and finally $b_0^{(2)}(\underline{EG}) = b_0^{(2)}(EG)$ by Theorem 4.14. Note that if G acted freely on X, we would get easily that G is finitely generated: by covering theory G would be a quotient of $\pi_1(G \backslash X)$ which is finitely generated because $G \backslash X$ has compact 2-skeleton. However, since we only assume that G acts properly, we need a more elaborate argument.

Lemma 4.22 *Suppose that for a group G there exists a nonempty, connected, proper, finite type G-CW complex X. Then G is finitely generated.*

Proof A connected CW complex is path-connected and any path in X connecting any two points in the 1-skeleton X_1 can be homotoped relative end points to a path inside X_1. Thus X_1 is a one-dimensional, connected CW complex with a proper, cellular and cocompact action by G. Since the action is cocompact, there exists a compact subcomplex $D \subset X$ such that the G-translates of D cover X. Since the action is proper, the set $S = \{g \in G: gD \cap D \neq \emptyset\}$ is finite. We claim that S generates G. So let $g \in G$. Pick a 0-cell $x_0 \in D$ and a finite chain $e_1, \ldots, e_n$ of oriented, closed 1-cells in X_1 joining x_0 to gx_0 so that the end point x_i of e_i is at the same time the initial point of e_{i+1} for $i = 1, \ldots, n-1$. Since the action is cellular, we can find group elements $g_i \in G$ such that $e_i \subset g_i D$. Enlarging D, if necessary, we can arrange that $g_1 = e$ and $g_n = g$. Since $x_i \in g_i D \cap g_{i+1} D$, we have $g_{i+1}^{-1} x_i \in D$ and $(g_i^{-1} g_{i+1}) g_{i+1}^{-1} x_i = g_i^{-1} x_i \in D$, whence $g_i^{-1} g_{i+1} \in S$. As $g = (g_1^{-1} g_2)(g_2^{-1} g_3) \cdots (g_{n-1}^{-1} g_n)$, the proof is complete. $\square$

Thus we are left with the task of computing $b_0^{(2)}(G)$ for a finitely generated group G. Let $S \subset G$ be a finite generating set. As explained above, we have a model for BG with $(BG)_1 \simeq \bigvee_{s \in S} S^1$. We observe that $(EG)_1$ is in this case the *Cayley graph* of G with respect to S. Indeed, the vertex set of $(EG)_1$ can be identified with the group G and two vertices $g_1, g_2 \in G$ are connected by an edge if and only if there exists $s \in S$ such that $sg_1 = g_2$. Accordingly, after picking a cellular basis, the first cellular differential is of the form

$$d_1: \bigoplus_{s \in S} \mathbb{Z}G \xrightarrow{\oplus_{s \in S} \cdot (s - e)} \mathbb{Z}G.$$

It is now convenient to consider the ℓ^2-cochain complex from Sect. 3.4 in Chap. 3 instead of the ℓ^2-chain complex we usually consider. The zeroth codifferential is given by

$$\mathrm{Hom}_{\mathbb{Z}G}(\mathbb{Z}G, \ell^2 G) \xrightarrow{\delta^0_{(2)}} \bigoplus_{s \in S} \mathrm{Hom}_{\mathbb{Z}G}(\mathbb{Z}G, \ell^2 G)$$

$$\varphi \mapsto \oplus_{s \in S} (x \mapsto \varphi(x(s - e))).$$

Thus a homomorphism $\varphi \in \ker \delta^0_{(2)}$ has the property that $\varphi(xs) = \varphi(x)$ for all $s \in S$. Since every element $g \in G$ is a finite word in the alphabet S, it follows $\varphi(xg) = \varphi(x)$ for all $g \in G$ and hence $g\varphi(e) = \varphi(g) = \varphi(e)$ for all $g \in G$. Writing $\varphi(e) = \sum_{g \in G} c_g g$, we see that this implies the coefficients c_g are constant throughout G. Since G is infinite, the ℓ^2-condition thus implies $c_g = 0$ for all $g \in G$ and hence $\varphi = 0$. From Theorem 3.24 we conclude

$$b_0^{(2)}(G) = b^0_{(2)}(EG) = \dim_{\mathcal{L}(G)} H^0_{(2)}(EG) = \dim_{\mathcal{L}(G)} \ker \delta^0_{(2)} = 0. \qquad \square$$

4.5.5 ℓ^2-Betti Numbers of Locally Compact Groups

Let G be a (second countable, Hausdorff) locally compact group. Up to scaling, there exists a unique nontrivial, countably additive measure μ on G, called the *Haar measure*, such that $\mu(gB) = \mu(B)$ for all $g \in G$ and all Borel subsets $B \subset G$. A discrete subgroup $\Gamma \leq G$ is called a *lattice* if μ induces a finite G-invariant measure on the quotient space G/Γ. For example, $\mathrm{PSL}(2, \mathbb{Z})$ is a lattice in $\mathrm{PSL}(2, \mathbb{R})$. It is a deep property based on the notion of ℓ^2-*Betti numbers of equivalence relations* that lattices $\Gamma, \Lambda \subset G$ in the same locally compact group satisfy *Gaboriau's proportionality principle* [55]:

$$\frac{b_n^{(2)}(\Gamma)}{\mu(G/\Gamma)} = \frac{b_n^{(2)}(\Lambda)}{\mu(G/\Lambda)}$$

for all $n \geq 0$. In particular, all lattices in G have vanishing and non-vanishing ℓ^2-Betti numbers in the same degree. The proportionality principle allows the definition $b_n^{(2)}(G, \mu) := \frac{b_n^{(2)}(\Gamma)}{\mu(G/\Gamma)}$ for the *n-th ℓ^2-Betti number of the locally compact group G* with a fixed (scaling of the) Haar measure μ provided G possesses a lattice $\Gamma \subset G$. It is not hard to see that locally compact groups which have lattices are *unimodular*, meaning that also $\mu(Bg) = \mu(B)$. More recently, a direct definition of $b_n^{(2)}(G, \mu)$ was given by H.D. Petersen for any unimodular locally compact group G, with or without lattice [142]. For groups with lattices, Gaboriau's machinery is then applied in [99]*Theorem B to verify the compatibility

$$b_n^{(2)}(\Gamma) = b_n^{(2)}(G, \mu) \cdot \mu(G/\Gamma).$$

Some example computations of ℓ^2-Betti numbers of locally compact groups can be found in [143].

Exercises

4.5.1 Recall our comment on p. 64 that for a homotopy equivalence $f : X \to X$ of a CW complex X the mapping torus $T(f)$ is (homotopy equivalent to) a *fibration* over S^1 (in the sense of *Serre*). This means that for any CW pair (Y, A) all *homotopy lifting problems*

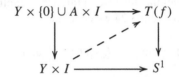

have a solution. Apply this fact to show that for every group G with finite type model for BG and for every $\varphi \in \mathrm{Aut}(G)$ we have $b_n^{(2)}(G \rtimes_\varphi \mathbb{Z}) = 0$ for all $n \geq 0$.

Chapter 5
Lück's Approximation Theorem

In Theorem 3.19 we saw that $\chi^{(2)}(X) = \chi(G \backslash X)$ for a finite, free G-CW complex X. Thus the alternating sum of ℓ^2-Betti numbers of X equals the alternating sum of ordinary Betti numbers of $G \backslash X$. One might wonder whether there is also some relation between the n-th ℓ^2-Betti number and the ordinary n-th Betti number by themselves.

5.1 The Statement

The example of the k-torus $\mathbb{T}^k$ illustrates that any such relation will have to be subtle as we have $b_n^{(2)}(\widetilde{\mathbb{T}^k}) = 0$ for all n while $b_n(\mathbb{T}^k) = \binom{k}{n}$. However, if X is a finite type H-CW complex for a finite group H, we saw in Example 3.14 that ordinary and ℓ^2-Betti numbers are related by the formula $b_n^{(2)}(X) = \frac{b_n(X)}{|H|}$. Given a proper, finite type G-CW complex X for a possibly infinite group G, every finite index normal subgroup $N \trianglelefteq G$ defines the G/N-CW complex $N \backslash X$ for which we thus have

$$b_n^{(2)}(G/N \curvearrowright N \backslash X) = \frac{b_n(N \backslash X)}{[G : N]}.$$

So one could hope to obtain the ℓ^2-Betti number $b_n^{(2)}(G \curvearrowright X)$ as a limit of the right hand side for "$N \to \{1\}$" whenever this expression is meaningful: G should have the property that finite index normal subgroups can come arbitrarily close to the trivial subgroup. The following definition makes this notion precise.

Definition 5.1 A *residual chain* in G is a sequence

$$G = G_0 \geq G_1 \geq G_2 \geq \cdots$$

© Springer Nature Switzerland AG 2019
H. Kammeyer, *Introduction to ℓ^2-invariants*, Lecture Notes in Mathematics 2247,
https://doi.org/10.1007/978-3-030-28297-4_5

of nested finite index normal subgroups $G_i \trianglelefteq G$ with $\bigcap_{i \geq 0} G_i = \{1\}$. A (countable) group G is called *residually finite* if it possesses a residual chain.

The class of residually finite group is reasonably large. It includes finite groups (trivial), free groups, finitely generated nilpotent groups, fundamental groups of 3-manifolds ([70] + geometrization) and, most notably, finitely generated *linear* groups: subgroups of $GL(n, K)$ for some n and some field K of arbitrary characteristic (see [135] for an account).

As a non-example, the Baumslag–Solitar groups

$$B(m, n) = \langle a, b \mid ba^m b^{-1} b^n \rangle$$

are not residually finite unless $|n| = 1$, $|m| = 1$, or $|n| = |m|$ as is proven in [127]. For some groups, residually finiteness fails in the strongest sense. Higman's group [74]

$$\langle a, b, c, d \mid a^{-1}bab^{-2}, b^{-1}cbc^{-2}, c^{-1}dcd^{-2}, d^{-1}ada^{-2} \rangle$$

is an infinite group with no finite quotients at all. Of course, infinite simple groups, for example *Tarski monsters*, are likewise not residually finite.

Theorem 5.2 (Lück, [115]) *Let X be a free, finite type G-CW complex. Assume G is residually finite and let (G_i) be any residual chain. Then for every $n \geq 0$ we have*

$$b_n^{(2)}(G \curvearrowright X) = \lim_{i \to \infty} \frac{b_n(G_i \backslash X)}{[G : G_i]}.$$

The theorem says that a positive n-th ℓ^2-Betti number detects "free homology growth": asymptotically, the free abelian rank of the homology group $H_n(G_i \backslash X)$ grows linearly in the index $[G : G_i]$ with speed $b_n^{(2)}(X)$. In view of Example 3.4, the covering space version of the theorem that was presented in the introduction as Theorem 1.6 is the special case of Theorem 5.2 when X is connected and simply-connected. The theorem obtains a purely group theoretic interpretation if X is moreover weakly contractible.

Theorem 5.3 *Let G be a residually finite group that has a finite type model for EG. Then for any residual chain (G_i) and every $n \geq 0$, we have*

$$b_n^{(2)}(G) = \lim_{i \to \infty} \frac{b_n(G_i)}{[G : G_i]}.$$

To conclude this result from Theorem 5.2, we only have to note that each subgroup G_i still acts freely on EG, so that $G_i \backslash EG$ is a model for BG_i. Again, if $b_n^{(2)}(G) > 0$, then $b_n(G_i) \to \infty$ for every residual chain.

In order to prove Theorem 5.2, we have to supplement our functional analytic toolbox from Chap. 2 by the beautiful theory of spectral calculus. This will occupy the next section.

Exercises

5.1.1 Let $G = F(a, b)$ be the free group on letters a and b. For $i \geq 1$, let $G_i \leq G$ be the subgroup given by

$$G_i = \langle a^i, a^k b a^{-k-1} : k = 0, \ldots, i-1 \rangle.$$

(i) Show that the subgroups G_i are nested, normal and of finite index in G.
(ii) Show that $\lim_{i \to \infty} [G : G_i] = \infty$ but that $\bigcap_{i \geq 1} G_i$ is nontrivial.
(iii) Verify by direct computation that nevertheless $b_1^{(2)}(G) = \lim_{i \to \infty} \frac{b_1(G_i)}{[G : G_i]}$.

5.2 Functional Calculus and the Spectral Theorem

Let $T \in B(H)$ be a bounded operator on a separable Hilbert space H. For what functions f can we define $f(T)$? Since $B(H)$ is a $\mathbb{C}$-algebra, we know what to do if $f \in \mathbb{C}[z]$ is a polynomial: for $f(z) = \sum_{k=0}^{n} a_k z^k$ we simply set $f(T) = \sum_{k=0}^{n} a_k T^k$. Similarly, if $f(z) = \sum_{k=0}^{\infty} a_k z^k$ is a power series which converges in some open disk $U = \{|z| < \|T\| + \varepsilon\}$, then $f(T) = \sum_{k=0}^{\infty} a_k T^k$ is still defined because the partial sums $\sum_{k=0}^{N} a_k T^k$ clearly form a Cauchy sequence in the Banach space $B(H)$. In fact, the condition that U should contain the closed disk around zero with radius $\|T\|$ is only a crude way to ensure the weaker condition that U contains the *spectrum* of T. It turns out that this is all we need to define $f(T)$.

Definition 5.4 The *spectrum* of T is the subset $\sigma(T) \subseteq \mathbb{C}$ given by

$$\sigma(T) = \{\zeta \in \mathbb{C} : \zeta \cdot \mathrm{id}_H - T \text{ is not bijective.}\}.$$

By Theorem 2.16, we could have equivalently required that $\zeta \cdot \mathrm{id}_H - T$ is not invertible. Since the set of invertible operators in $B(H)$ is open, $\sigma(T)$ is a closed subset of $\mathbb{C}$. We have Gelfand's *spectral radius formula*

$$r(T) := \sup_{\zeta \in \sigma(T)} |\zeta| = \lim_{n \to \infty} \|T^n\|^{\frac{1}{n}}$$

and in particular $r(T) \leq \|T\|$ which shows that $\sigma(T)$ is bounded, hence compact. The complement $\varrho(T) = \mathbb{C} \setminus \sigma(T)$ is also known as the *resolvent set* of T. By the inverse mapping theorem (Theorem 2.16), each $\zeta \in \varrho(T)$ defines a bounded operator $(\zeta \cdot \mathrm{id}_H - T)^{-1}$ so that we have the *resolvent mapping*

$$R(T) \colon \varrho(T) \to B(H), \quad \zeta \mapsto \frac{1}{\zeta - T}.$$

If $\sigma(T)$ was empty, then for every $x, y \in H$, the inner product $\langle R(T)(\zeta)x, y \rangle$ would define an entire function tending to zero for $\zeta \to \infty$. By Liouville's theorem, we then must have $\langle R(T)x, y \rangle = 0$ for all $x, y \in H$, thus $R(T) = 0$ which is absurd. Thus $\sigma(T)$ is always nonempty. Conversely, any compact nonempty subset of $\mathbb{C}$ occurs as spectrum of a bounded operator in $B(H)$ as you will prove in the guided Exercise 5.2.1.

Now the key observation to define $f(T)$ for a power series f converging in a neighborhood U of $\sigma(T)$ is that f defines a *holomorphic function* $f \colon U \to \mathbb{C}$ for which we have the *Cauchy formula*

$$f(z) = \frac{1}{2\pi \mathrm{i}} \int_\gamma \frac{f(\zeta)}{\zeta - z} \, \mathrm{d}\zeta$$

where γ is any (piecewise) smooth closed curve in U winding once around z. It is possible to define "operator valued integration" by mimicking the classical definition in terms of Riemann sums of finer and finer partitions, only that convergence is now required with respect to the operator norm of $B(H)$. Let Γ be a finite set of closed curves in U such that the *inner points* I_Γ, those $z \in \mathbb{C}$ that have winding number one with respect to Γ, satisfy

$$\sigma(T) \subset I_\Gamma \subset U.$$

Such a set Γ always exists because $\sigma(T)$ is compact. Then

$$f(T) = \frac{1}{2\pi \mathrm{i}} \sum_{\gamma \in \Gamma} \int_\gamma \frac{f(\zeta)}{\zeta - T} \, \mathrm{d}\zeta$$

gives a well-defined bounded operator $f(T) \in B(H)$ satisfying the *spectral mapping theorem* $\sigma(f(T)) = f(\sigma(T))$. The construction gives the existence part of the following theorem.

Theorem 5.5 (Holomorphic Functional Calculus) *Let $T \in B(H)$ and let U be an open neighborhood of $\sigma(T)$. Then there is a unique homomorphism*

$$O(U) = \{f \colon U \to \mathbb{C} \text{ holomorphic}\} \longrightarrow B(H), \quad f \mapsto f(T)$$

of $\mathbb{C}$-algebras which is unit preserving, $\chi_U(T) = \mathrm{id}_H$, satisfies $\mathrm{id}_U(T) = T$, and is continuous with respect to uniform convergence on compact sets in U.

Recall that χ_U denotes the characteristic function of the set U. Uniqueness is easy: such a homomorphism is determined on polynomials and holomorphic functions on U can be identified with convergent power series on U. These are uniform limits of the partial sums on any compact subset of U. Holomorphic functional calculus is in general not an injective homomorphism. Nonetheless, the identity theorem says it is injective if U is connected and $\sigma(T)$ has a cluster point.

We remark that for the entire construction, it was not important that we were working in $B(H)$. Any unital *Banach algebra A* would have worked equally fine: an associative $\mathbb{C}$-algebra A with $1 \in A$, endowed with a complete norm $\| \cdot \|$ satisfying $\|xy\| \leq \|x\| \|y\|$ for all $x, y \in A$. On $B(H)$, however, we have the additional structure of a $*$-operation and if $T = T^*$ is self-adjoint, one can see with the help of Exercise 2.2.6 (ii) that $\sigma(T) \subset \mathbb{R}$. Moreover, the spectral radius formula reduces in this case to $r(T) = \|T\|$ (Exercise 5.2.2). For these operators, we can improve the domain of definition from $O(U)$ to the C^*-algebra $C(\sigma(T), \mathbb{C})$ of continuous $\mathbb{C}$-valued functions on the compact, nonempty set $\sigma(T)$ with $*$-operation given by complex conjugation.

Theorem 5.6 (Continuous Functional Calculus) *Let $T \in B(H)$ be self-adjoint. Then there is a unique isometric $*$-embedding of C^*-algebras*

$$C(\sigma(T), \mathbb{C}) \longrightarrow B(H), \quad f \mapsto f(T)$$

which is unit preserving, $\chi_{\sigma(T)}(T) = \mathrm{id}_H$, and satisfies $\mathrm{id}_{\sigma(T)}(T) = T$.

Proof (Idea) Requiring $\chi_{\sigma(T)}(T) = \mathrm{id}_H$ and $\mathrm{id}_{\sigma(T)}(T) = T$ implies that for a polynomial p, the operator $p(T)$ is given by evaluating p in the operator T. The core part of the proof is to show that this is norm preserving, meaning $\|p(T)\| = \|p\|$ where the latter denotes the sup-norm on $C(\sigma(T), \mathbb{C})$. The polynomials on $\sigma(T)$ form a point separating subalgebra in $C(\sigma(T), \mathbb{C})$ so that the unital $*$-algebra they generate is dense by the Stone–Weierstraß theorem. It follows that the map $p \mapsto p(T)$ has a unique extension to a continuous $*$-homomorphism on $C(\sigma(T), \mathbb{C})$ which is clearly norm-preserving, hence injective. □

The spectral mapping theorem $\sigma(f(T)) = f(\sigma(T))$ also holds for continuous calculus. If $\lambda \in \sigma(T)$ is an eigenvalue so that there is $x \in H$ nonzero with $Tx = \lambda x$, then $f(T)x = f(\lambda)x$. This is immediate from continuity when approximating f by polynomials. Preservation of $*$-operation and the spectral mapping theorem show that precisely the real valued functions $f \in C(\sigma(T), \mathbb{R}) \subset C(\sigma(T), \mathbb{C})$ give a self-adjoint operator $f(T) = f(T)^*$.

Proposition 5.7 *The operator $f(T)$ is positive if and only if $f \geq 0$.*

Proof In one direction, this follows because for $f \in C(\sigma(T), \mathbb{R}_{\geq 0})$ we get $f(T) = \sqrt{f}(T)^* \sqrt{f}(T) \geq 0$. In the other direction, if $f \in C(\sigma(T), \mathbb{R})$ has $f(\lambda) < 0$ for some $\lambda \in \sigma(T)$, use *Weyl's criterion*: there are $x_n \in H$ with $\|x_n\| = 1$ and

$\lim_{n\to\infty} \|(T-\lambda)x_n\| = 0$. Hence

$$\langle f(T)x_n, x_n \rangle = \langle (f(T) - f(\lambda))x_n, x_n \rangle + f(\lambda)$$

is negative for big enough n by Exercise 5.2.4.   □

The proposition says in particular that for self-adjoint $T \in B(H)$ and any fixed vector $x \in H$, the linear functional

$$\Phi_{x,T} \colon C(\sigma(T), \mathbb{C}) \longrightarrow \mathbb{C}, \quad f \mapsto \langle x, f(T)x \rangle$$

is *positive*: if $f \in C(\sigma(T), \mathbb{R}_{\geq 0})$, then $\Phi_{x,T}(f) \geq 0$. This observation will lead us to yet another extension of the domain for functional calculus.

Theorem 5.8 (Riesz Representation Theorem) *Let X be a compact Hausdorff space and let $\Phi \colon C(X, \mathbb{C}) \to \mathbb{C}$ be a positive linear functional. Then there exists a unique regular Borel measure μ on X such that*

$$\Phi(f) = \int f \, d\mu$$

for all $f \in C(X, \mathbb{C})$. The total mass of μ is given by $\mu(X) = \|\Phi\|$.

The reader will find the arduous proof in [153, Theorem 2.14, p. 40].

Definition 5.9 The *spectral measure* of T associated with x is the unique measure $\mu_{x,T}$ representing the positive linear functional $\Phi_{x,T}$.

We thus have

$$\langle x, f(T)x \rangle = \int f \, d\mu_{x,T} \tag{5.1}$$

for all $f \in C(\sigma(T), \mathbb{C})$. Now the decisive observation is that the right hand side is in fact defined for all f from $\mathcal{B}(\sigma(T), \mathbb{C})$, the bounded complex-valued Borel measurable functions on $\sigma(T)$. Thus we can simply define the values $\langle x, f(T)x \rangle \in \mathbb{C}$ for $x \in H$ and $f \in \mathcal{B}(\sigma(T), \mathbb{C})$ by (5.1). Then polarization (recall Exercise 2.2.4) determines the values $\langle x, f(T)y \rangle$ for all $x, y \in H$. The Riesz lemma (Theorem 2.18) provides for each $x \in H$ a unique vector $z \in H$ such that $\langle x, f(T)y \rangle = \langle z, y \rangle$ for all $y \in H$. This determines the operator $f(T)^*$ and hence $f(T)$ for $f \in \mathcal{B}(\sigma(T), \mathbb{C})$.

Theorem 5.10 (Borel Functional Calculus) *Let $T \in B(H)$ be self-adjoint. Continuous functional calculus extends uniquely to a continuous $*$-homomorphism*

$$\mathcal{B}(\sigma(T), \mathbb{C}) \longrightarrow B(H), \quad f \mapsto f(T).$$

Continuity of the $*$-homomorphism can more precisely be stated as $\|f(T)\| \le \|f\|$ where $\|f\|$ is again the sup-norm. This inequality becomes an equality once we identify two bounded Borel functions if they agree $\mu_{x,T}$-almost everywhere for all $x \in H$. In fact, there is an up to equivalence unique *basic measure* μ_T on $\sigma(T)$ whose null sets are precisely the measurable subsets of $\sigma(T)$ which are $\mu_{x,T}$-null sets for all $x \in H$ [33, Proposition 4 (iii), p. 130]. Hence under the above identification, $\mathscr{B}(\sigma(T), \mathbb{C})$ is turned into $L^\infty(\sigma(T), \mu_T)$, the μ_T-essentially bounded Borel measurable complex valued functions on $\sigma(T)$ up to equality μ_T-almost everywhere. Borel functional calculus then defines an isometric $*$-embedding

$$L^\infty(\sigma(T), \mu_T) \longrightarrow B(H)$$

which is not only an embedding as C^*-algebra, but in fact an ultraweakly continuous and weakly closed embedding as von Neumann algebra [33, Proposition 1, p. 128]. Moreover:

Proposition 5.11 *If a sequence of functions $f_n \in L^\infty(\sigma(T), \mu_T)$ converges μ_T-almost everywhere to some $f \in L^\infty(\sigma(T), \mu_T)$, then the sequence of operators $f_n(T)$ converges strongly to $f(T)$.*

Proof The weak convergence follows from (5.1) and the bounded convergence theorem. For strong convergence, we need additionally that $\|f_n(T)x\|$ converges to $\|f(T)x\|$ for all $x \in H$. But that is equivalent to the weak convergence of $|f_n|^2(T)$ to $|f|^2(T)$, so we are done. $\qquad\square$

The next result says the basic measure μ_T has atoms precisely at the eigenvalues of T.

Proposition 5.12 *An element $\lambda \in \sigma(T)$ is an eigenvalue of T with normalized eigenvector $x \in H$ if and only if $\mu_{x,T}$ is the Dirac measure at λ.*

Proof Let δ_λ denote the Dirac probability measure with support $\{\lambda\}$. If $x \in H$ with $\|x\| = 1$ is an eigenvector of T with eigenvalue $\lambda \in \sigma(T)$, then (5.1) says that $\int f \, d\delta_\lambda = \int f \, d\mu_{x,T}$ for all continuous functions f on $\sigma(T)$. But any characteristic function χ_A of a closed set $A \subset \sigma(T)$ is a monotone limit of continuous functions and the closed sets form a $\cap$-stable generating system of the σ-algebra of Borel sets. This implies $\mu_{x,T} = \delta_\lambda$ (see also [12, Lemma 30.14, p. 231]). Conversely, if $\mu_{x,T} = \delta_\lambda$, then

$$\|(\lambda \cdot \mathrm{id}_H - T)x\|^2 = \lambda^2 - 2\lambda\langle x, Tx \rangle + \langle x, T^2 x \rangle$$

$$= \lambda^2 - 2\lambda \int s \, d\mu_{x,T}(s) + \int s^2 \, d\mu_{x,T}(s)$$

$$= \lambda^2 - 2\lambda^2 + \lambda^2 = 0,$$

so x is an eigenvector of T with eigenvalue λ and

$$\|x\|^2 = \langle x, \mathrm{id}_H(x)\rangle = \langle x, \chi_{\sigma(T)}(T)x\rangle = \int \chi_{\sigma(T)}\,d\mu_{x,T} = \delta_\lambda(\sigma(T)) = 1,$$

so x is normalized.   □

Proposition 5.13 *If* $\lambda \in \sigma(T)$ *is an eigenvalue with eigenvector* $x \in H$, *then for every* $f \in L^\infty(\sigma(T), \mu_T)$, *we have* $f(T)x = f(\lambda)x$.

Proof First note that by the last proposition, the value $f(\lambda)$ is well-defined. We already know the statement holds true if f is continuous. In view of Proposition 5.11, it remains to show that for every function $f \in \mathscr{B}(\sigma(T), \mathbb{C})$, there exists a sequence from the subalgebra $C(\sigma(T), \mathbb{C})$ of continuous functions which converges pointwise to f μ_T-almost everywhere. To construct such a sequence, we can start with a sequence $f_n \in C(\sigma(T), \mathbb{C})$ which converges to f in L^1-norm, meaning $\lim_{n\to\infty} \int |f - f_n|\,d\mu_T = 0$. Such a sequence exists because f is an L^1-limit of simple functions which in turn have L^1-approximations by continuous functions. It then follows from basic measure theory that f_n converges in measure to f and consequently, a subsequence converges to f μ_T-almost everywhere.   □

Of course it does not even make sense to ask if the spectral theorem $\sigma(f(T)) = f(\sigma(T))$ is true for Borel calculus. We remark that both the continuous and the Borel functional calculus extend from self-adjoint to normal operators $T \in B(H)$, essentially because these operators still generate abelian C^*- and von Neumann algebras. Table 5.1 gives an overview of the three different types of functional calculus we have described. As the functions—from holomorphic via continuous to measurable—become more and more general, the ranges become bigger and bigger operator algebras.

Remember that for the holomorphic functional calculus, the operator T can lie in any unital Banach algebra. Similarly, for the continuous functional calculus, T may lie in any unital C^*-algebra and for the Borel functional calculus, T may lie in any von Neumann algebra on a separable Hilbert space. The ranges will then be norm and weakly closed subalgebras, respectively.

The feature that C^*-algebras come with a continuous functional calculus while von Neumann algebras have a measurable functional calculus is yet another striking

Table 5.1 Various flavors of functional calculus

Functional calculus	$T \in B(H)$	Domain	Range
Holomorphic	Any	$O(U)$	Subalgebra of unital Banach algebra generated by T
Continuous	Normal	$C(\sigma(T), \mathbb{C})$	Unital C^*-algebra generated by T
Borel	Normal	$L^\infty(\sigma(T), \mu_T)$	von Neumann algebra generated by T

corroboration of the philosophy we alluded to in Remark 2.27: C^*-algebras are non-commutative topological spaces and von Neumann algebras are noncommutative measure spaces.

The decisive advantage of the passage from the abelian C^*-algebra $C(\sigma(T), \mathbb{C})$ to the abelian von Neumann algebra $L^\infty(\sigma(T), \mathbb{C})$, is that the latter contains (and is actually generated by) the characteristic functions χ_A for measurable subsets $A \subset \sigma(T)$. Since $\chi_A^2 = \chi_A = \overline{\chi_A}$, these are orthogonal projections in $L^\infty(\sigma(T), \mathbb{C})$. Hence so are the corresponding operators $P_T(A) := \chi_A(T)$ in $B(H)$, called the *spectral projections* of T. They form a *projection valued measure*: we have $P_T(\sigma(T))) = \mathrm{id}_H$ and every $x \in H$ gives a Borel measure $A \mapsto \langle x, P_T(A)x \rangle$ on $\sigma(T)$, namely the spectral measure $\mu_{x,T}$. Integration of bounded measurable functions can be defined with respect to projection valued measures in the usual way. The result is a bounded operator and Borel functional calculus takes the elegant form

$$f(T) = \int f \, dP_T.$$

By (5.1), the projection valued measure P_T and the spectral measure $\langle x, P_T x \rangle = \mu_{x,T}$ satisfy the compatibility relation

$$\left\langle x, \int f \, dP_T \, x \right\rangle = \int f \, d\langle x, P_T x \rangle.$$

We spell out the particular case $f = \mathrm{id}_{\sigma(T)}$.

Theorem 5.14 (Spectral Theorem) *Let $T \in B(H)$ be self-adjoint. Then*

$$T = \int_{\sigma(T)} \lambda \, dP_T(\lambda).$$

The theorem is a vast generalization of the fact from linear algebra that a Hermitian matrix is (unitarily) diagonalizable with real eigenvalues. We have indeed the following observation.

Proposition 5.15 *If $\lambda \in \sigma(T)$ is an eigenvalue, then $P_T(\{\lambda\})$ is the orthogonal projection onto the eigenspace of λ.*

Proof Let $x \in H$ be an eigenvector for λ. By Proposition 5.13, we obtain

$$P_T(\{\lambda\}) x = \chi_{\{\lambda\}}(T) x = \chi_{\{\lambda\}}(\lambda) x = x,$$

hence x lies in the image of $P_T(\{\lambda\})$. Conversely, let x be a nonzero vector in the image of $P_T(\{\lambda\})$. Setting $\{\lambda\}^c = \sigma(T) \setminus \{\lambda\}$, we obtain

$$\chi_{\{\lambda\}^c}(T) x = \chi_{\{\lambda\}^c}(T)\chi_{\{\lambda\}}(T) x = (\chi_{\{\lambda\}^c} \cdot \chi_{\{\lambda\}})(T) x = 0.$$

Therefore

$$Tx = \mathrm{id}_{\sigma(T)}(T)x = (\lambda\, \chi_{\{\lambda\}} + \mathrm{id}_{\sigma(T)}\, \chi_{\{\lambda\}^c})(T)x = \lambda x + T\, \chi_{\{\lambda\}^c}(T)x = \lambda x,$$

so x is an eigenvector of λ.   □

This proposition concludes our excursion to spectral calculus. We are now suitably armed to attack the proof of Lück's approximation theorem.

Exercises

5.2.1 Let $D \subset \mathbb{C}$ be a compact, nonempty subset.

(i) Construct a countable, dense subset $X \subset D$.
(ii) Show that $T\colon L^2(X, \mu) \to L^2(X, \mu)$, $f(x) \mapsto xf(x)$ defines a bounded operator if μ denotes the counting measure on X. What is $\|T\|$?
(iii) Show that $\sigma(T) = D$.

5.2.2 Let $T \in B(H)$. Show that $r(T) = \|T\|$ if

(i) T is self-adjoint,
(ii) T is normal.

Hint: Use the C^*-identity $\|T^*T\| = \|T\|^2$.

5.2.3 Gelfand's spectral radius formula is not only true in $B(H)$ but actually in any unital Banach algebra A. Conclude that if A is even a C^*-algebra, then the unital $*$-algebra structure of A determines the norm. Thus the unital $*$-algebra structure of a C^*-algebra determines the topology!

5.2.4 Let $T \in B(H)$ be self-adjoint. *Weyl's criterion* says that $\lambda \in \sigma(T)$ if and only if λ is an *approximate eigenvalue*, meaning λ has an *approximate eigenvector*: a sequence $(x_n) \subset H$ with $\|x_n\| = 1$ and $\lim_n \|(T - \lambda)x_n\| = 0$. In that case, show that for every $f \in C(\sigma(T), \mathbb{C})$, the sequence (x_n) is an approximate eigenvector of $f(T)$ with approximate eigenvalue $f(\lambda)$. *Hint: First assume f is a polynomial.*

5.2.5 Let X be a compact Hausdorff space. Find all projections in the abelian C^*-algebra $C(X, \mathbb{C})$.

5.3 The Proof

In view of Example 3.14, Lück's approximation theorem asserts that

$$b_n^{(2)}(G \curvearrowright X) = \lim_{i \to \infty} b_n^{(2)}(G/G_i \curvearrowright G_i \backslash X) \tag{5.2}$$

for a free, finite-type G-CW complex X, any residual chain (G_i) in G, and each fixed $n \geq 0$. As the first step of the proof, we translate this topological statement to an algebraic one. We fix a cellular basis (p. 40) of X and obtain cellular bases for all the (G/G_i)-CW complexes $G_i \backslash X$ by composing with the canonical projections $X \to G_i \backslash X$. These define identifications

$$C_n^{(2)}(X) \cong (\ell^2 G)^k \quad \text{and} \quad C_n^{(2)}(G_i \backslash X) \cong (\ell^2(G/G_i))^k$$

where $k = k(n)$ is the number of equivariant n-cells in X. Under this identification, the ℓ^2-Laplacian $\Delta_n^{(2)}$ of X from p. 55 acts on $(\ell^2 G)^k$ by right multiplication with a $*$-invariant matrix $D \in M(k, k; \mathbb{Z}G)$ because the $(n + 1)$-skeleton of X is finite. Correspondingly, the n-th ℓ^2-Laplacian of $G_i \backslash X$ acts on $(\ell^2(G/G_i))^k$ by right multiplication with the matrix $D_i \in M(k, k; \mathbb{Z}(G/G_i))$ obtained from D by applying the canonical $*$-ring homomorphism $\mathbb{Z}G \to \mathbb{Z}(G/G_i)$ to the entries. Because of Proposition 3.23, in these terms Lück's theorem takes the form

$$\dim_{\mathcal{R}(G)} \ker(\cdot D) = \lim_{i \to \infty} \dim_{\mathcal{R}(G/G_i)} \ker(\cdot D_i).$$

In the next step, we exploit our excursion to functional calculus to translate this algebraic statement into a measure theoretic one. To this end, let $\varepsilon = e \oplus \cdots \oplus e$ be the diagonal vector in $(\ell^2 G)^k$ consisting of the unit vector $e \in \ell^2 G$ in each of the k coordinates. Then Proposition 5.15 gives

$$\dim_{\mathcal{R}(G)} \ker(\cdot D) = \operatorname{tr}_{\mathcal{R}(G)} P_{\cdot D}(\{0\}) = \langle \varepsilon, P_{\cdot D}(\{0\})\varepsilon \rangle = \mu(\{0\})$$

where $P_{\cdot D}$ is the projection valued measure and $\mu := \mu_{\varepsilon, \cdot D}$ is the spectral measure of the operator $\cdot D$ associated with ε. Similarly, we obtain

$$\dim_{\mathcal{R}(G/G_i)} \ker(\cdot D_i) = \mu_i(\{0\})$$

where $\mu_i := \mu_{\varepsilon_i, \cdot D_i}$ is the spectral measure of the operator $\cdot D_i$ associated with the vector $\varepsilon_i = G_i \oplus \cdots \oplus G_i \in (\ell^2(G/G_i))^k$ consisting of the unit element $G_i \in G/G_i$ in each of the k coordinates. Thus Lück's approximation theorem ultimately asserts a convergence property of spectral measures, to wit

$$\mu(\{0\}) = \lim_{i \to \infty} \mu_i(\{0\}). \tag{5.3}$$

It is in this formulation that the theorem becomes accessible because there is a good deal of techniques to investigate convergence questions for measures. We start by showing that the sequence of measures μ_i converges *weakly* to μ. To do so, recall from the end of the proof of Proposition 3.8 that we have

$$\| \cdot D\| \leq k^2 \cdot \|D\|_1 =: d.$$

The same bound works for the reduced matrices, $\| \cdot D_i \| \leq d$ for all i, so that we can consider μ_i and μ as measures on the closed interval $[0, d]$.

Proposition 5.16 *For all continuous functions* $f \in C([0, d], \mathbb{R})$ *we have*

$$\int f \, d\mu = \lim_{i \to \infty} \int f \, d\mu_i.$$

Proof First assume f is a polynomial with real coefficients. Then $\int f \, d\mu = \operatorname{tr}_{\mathcal{R}(G)}(\cdot f(D))$ is the sum of the coefficients of the unit element e in the diagonal entries of the matrix $f(D) \in M(k, k; \mathbb{R}G)$. Similarly, $\int f \, d\mu_i = \operatorname{tr}_{\mathcal{R}(G/G_i)}(\cdot f(D_i))$ is the sum of the coefficients of the unit element G_i in the diagonal entries of $f(D_i) \in M(k, k; \mathbb{R}(G/G_i))$. Observe that $f(D_i)$ is still obtained from $f(D)$ by applying the ring homomorphism $\mathbb{R}G \to \mathbb{R}(G/G_i)$ to the entries. Thus if N is so large that we have $g \notin G_N$ for all those $g \in G$ that have a nonzero coefficient in any of the diagonal entries of $f(D)$, then $\operatorname{tr}_{\mathcal{R}(G/G_i)}(\cdot f(D_i)) = \operatorname{tr}_{\mathcal{R}(G)}(\cdot f(D))$ for all $i \geq N$. In particular, we obtain the asserted convergence.

In the general case $f \in C([0, d], \mathbb{R})$, we know that f is a uniform limit of real polynomials by the Stone–Weierstrass theorem and the assertion follows from the bounded convergence theorem. $\square$

What this result has to say on whether (5.3) holds true, is captured by the following classical theorem of measure theory [43, Theorem 4.10, p. 385].

Theorem 5.17 (Portmanteau Theorem) *Let E be a compact metric space and let v_i and v be finite Borel measures on E with the same total mass. Then the following are equivalent.*

(i) *For all $f \in C(E, \mathbb{R})$, we have $\lim_{i \to \infty} \int f \, dv_i = \int f \, dv$.*
(ii) *For all closed sets $A \subset E$, we have $\limsup_{i \to \infty} v_i(A) \leq v(A)$.*
(iii) *For all open sets $U \subset E$, we have $\liminf_{i \to \infty} v_i(U) \geq v(U)$.*

Note that our spectral measures μ_i and μ all have total mass k. Combining Proposition 5.16 with the "(i) $\Rightarrow$ (ii)" part of the theorem, we obtain $\limsup_{i \to \infty} \mu_i(\{0\}) \leq \mu(\{0\})$, in other words

$$b_n^{(2)}(G \curvearrowright X) \geq \limsup_{i \to \infty} b_n^{(2)}(G/G_i \curvearrowright G_i \backslash X),$$

which is half of what we are striving for. This inequality is sometimes known as *Kazhdan's inequality*. It already says Lück's theorem holds true if $b_n^{(2)}(X) = 0$. For the general case, however, we still need to know that also

$$\liminf_{i \to \infty} \mu_i(\{0\}) \geq \mu(\{0\}). \tag{5.4}$$

This is not automatic from weak convergence: The Dirac measures $\delta_{1/i}$ on the measurable space $[0, d]$ converge weakly to δ_0 by continuity of $f \in C([0, d], \mathbb{R})$ at

Fig. 5.1 A logarithmic
bound. The figure also shows
the graph of the distribution
function of the measure $\delta_{1/i}$
which would violate any such
bound for large i

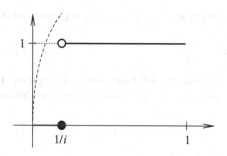

zero. But $\delta_{1/i}(\{0\}) = 0$ for all $i \geq 1$ whereas $\delta_0(\{0\}) = 1$. The measures $\delta_{1/i}$ would
occur as our measures μ_i if for instance the matrices D_i were the constant diagonal
matrices with entries $1/i \cdot G_i$. But there is a simple reason why no scenario of this
sort can occur: Our matrices D_i have entries from the group ring $\mathbb{Z}(G/G_i)$ which
has *integral* coefficients!

Lück's decisive insight was that this integrality leads to the existence of a
continuous function, independent of the residual chain (G_i), that uniformly bounds
all the *positive spectral distribution functions* $\lambda \mapsto \mu_i((0, \lambda))$ but still tends to zero
for small positive λ. We indicate such a function as the dashed plot in Fig. 5.1.

Proposition 5.18 (Logarithmic Bound) *For all i and $\lambda \in (0, 1)$ we get*

$$\mu_i((0, \lambda)) \leq \frac{k \log d}{|\log \lambda|}.$$

Proof We agree to fix $i \geq 0$ and $\lambda \in (0, 1)$ throughout the proof. Setting
$r = k\,[G : G_i]$, we can consider D_i as a symmetric $(r \times r)$-matrix with coefficients
in $\mathbb{Z}$ operating on $\mathbb{C}^r$ by multiplication. Let $\lambda_1 < \cdots < \lambda_s$ be the distinct
positive eigenvalues of D_i with multiplicities $m_1, \ldots, m_s$. Proposition 5.15 and
Example 3.14 show that for each $j = 1, \ldots, s$, we have

$$\mu_i(\{\lambda_j\}) = \mathrm{tr}_{\mathcal{R}(G/G_i)}\, P._{D_i}(\{\lambda_j\}) = \dim_{\mathcal{R}(G/G_i)}\left(\mathrm{im}\, P._{D_i}(\{\lambda_j\})\right) = \frac{m_j}{[G : G_i]}.$$

Say the first t eigenvalues λ_j are strictly smaller than λ. Since μ_i is supported on
$\sigma(\cdot D_i)$, which either equals $\{0, \lambda_1, \ldots, \lambda_s\}$ or $\{\lambda_1, \ldots, \lambda_s\}$, we obtain

$$\mu_i((0, \lambda)) = \sum_{j=1}^{t} \mu_i(\{\lambda_j\}) = \frac{m_1 + \cdots + m_t}{[G : G_i]}.$$

The characteristic polynomial p of D_i satisfies the relation

$$\frac{p(x)}{x^{r-R}} = (\lambda_1 - x)^{m_1} \cdots (\lambda_s - x)^{m_s}.$$

where $R = m_1 + \cdots + m_s$ is the rank of the matrix D_i. Setting $x = 0$ gives

$$1 \leq \lambda_1^{m_1} \cdots \lambda_s^{m_s} \tag{5.5}$$

because the left hand side is a polynomial with integer coefficients and a positive integer is at least one. From this we obtain the estimate

$$1 \leq \lambda_1^{m_1} \cdots \lambda_t^{m_t} \cdot \lambda_{t+1}^{m_{t+1}} \cdots \lambda_s^{m_s} \leq \lambda^{m_1 + \cdots + m_t} \, d^r$$

by the spectral radius formula and again because $\| \cdot D_i \| \leq d$ for all i. Taking logarithm and keeping in mind that $\log \lambda < 0$, this is equivalent to

$$\frac{m_1 + \cdots + m_t}{[G : G_i]} \leq \frac{k \log d}{|\log \lambda|}. \qquad \square$$

With this proposition at our disposal, we can easily finish the proof. For all $\lambda \in (0, 1)$, we have $\mu_i(\{0\}) = \mu_i([0, \lambda)) - \mu_i((0, \lambda))$ so that Proposition 5.16, the "(i) $\Rightarrow$ (iii)" part of Theorem 5.17, and Proposition 5.18 give

$$\liminf_{i \to \infty} \mu_i(\{0\}) \geq \mu([0, \lambda)) - \frac{k \log d}{|\log \lambda|}.$$

Since this holds for arbitrary $\lambda \in (0, 1)$, we also have

$$\liminf_{i \to \infty} \mu_i(\{0\}) \geq \lim_{\lambda \to 0^+} \left(\mu([0, \lambda)) - \frac{k \log d}{|\log \lambda|} \right) = \inf_{\lambda > 0} \mu([0, \lambda)) \geq \mu(\{0\}).$$

We thus verified (5.4) and the proof of Lück's approximation theorem is complete.

Exercises

5.3.1 Review the proof and point your finger to where exactly the various assumptions of Lück's approximation theorem enter: X of *finite type*, X is *free*, *normal* subgroups, subgroups with *trivial total intersection*, *finite index* subgroups, *nested subgroups*.

5.4 Extensions

In this section, we want to take Lück's approximation theorem to the limit and discuss for each of its assumptions in how far they are necessary or allow for generalization. Towards the end, we report on some recent variants of the approximation theorem, illustrating that this result keeps inspiring researchers to this day.

5.4.1 Infinite Type G-CW Complexes

The proof of Lück's approximation theorem in the previous section heavily relies on the observation that ℓ^2-Laplacians are realized by matrices over the group ring with a finite number of rows and columns. Thus it appears unpromising to try and loosen the finite type assumption for the G-CW complex X. Indeed, Lück and Osin's group Q from the end of Sect. 4.4 is a finitely generated, residually finite, infinite p-group with positive first ℓ^2-Betti number. Since torsion groups have vanishing first Betti number, the group Q violates the conclusion of Corollary 5.3. So one cannot even extend the approximation theorem for the first ℓ^2-Betti number from finitely presented to finitely generated groups.

5.4.2 Proper G-CW Complexes

In this paragraph we weaken the assumption on the action of G on X from being free to being proper. If the group G has a finite type model for EG, it is clear that this can be done by the Borel construction: we can apply Lück's theorem to the free finite type G-CW complex $EG \times X$ which has the same ℓ^2-Betti numbers as the proper finite type G-CW complex X. In general, however, we will have to adapt the arguments of the preceding section to the occurrence of stabilizer groups.

Theorem 5.19 *Let X be a proper, finite type G-CW complex. Assume G is residually finite and let (G_i) be any residual chain. Then for every $n \geq 0$ we have*

$$b_n^{(2)}(G \curvearrowright X) = \lim_{i \to \infty} \frac{b_n(G_i \backslash X)}{[G : G_i]}.$$

Proof We can still factor out the normal subgroups $G_i \trianglelefteq G$ to obtain the G/G_i-CW complexes $G_i \backslash X$. Each equivariant n-cell $G/H \times D^n$ in X with finite stabilizer group $H \leq G$ corresponds to an equivariant quotient cell $G/HG_i \times D^n$ in the G/G_i-CW complex $G_i \backslash X$ with finite stabilizer group $H/H \cap G_i \cong HG_i/G_i \leq G/G_i$. In particular, the G/G_i-CW complexes $G_i \backslash X$ are proper. By Example 3.14 we again have the reformulation (5.2) of the approximation theorem.

We pick a cellular basis of X. This realizes the ℓ^2-Laplacian $\Delta_n^{(2)}$ of X as an operator on $\bigoplus_{r \in I_n} \ell^2(G/H_r)$. Proposition 3.8 explains that this operator is given by right multiplication with the $(k \times k)$-matrix $D_{rs} \in \mathbb{Z}(G/H_s)^{H_r}$ by the well-defined rule $gH_r \cdot D_{rs} = gD_{rs}$. Here $k = |I_n|$ is again the number of equivariant n-cells.

For each i, we have canonical $\mathbb{Z}$-module homomorphisms

$$p_i : \mathbb{Z}(G/H_s)^{H_r} \longrightarrow \mathbb{Z}(G/H_s G_i)^{H_r/H_r \cap G_i},$$

and the n-th ℓ^2-Laplacian of $G_i \backslash X$ acts on $\bigoplus_{r \in I_n} \ell^2(G/H_r G_i)$ by right multiplication with the matrix $(D_i)_{rs} = p_i(D_{rs})$. Consider the vectors

$$\varepsilon = \frac{H_1}{\sqrt{|H_1|}} \oplus \cdots \oplus \frac{H_k}{\sqrt{|H_k|}} \in \bigoplus_{r \in I_n} \ell^2(G/H_r) \quad \text{and}$$

$$\varepsilon_i = \frac{H_1 G_i}{\sqrt{|H_1|}} \oplus \cdots \oplus \frac{H_k G_i}{\sqrt{|H_k|}} \in \bigoplus_{r \in I_n} \ell^2(G/H_r G_i).$$

As in the preceding section, these allow the reformulation of the theorem as

$$\mu(\{0\}) = \lim_{i \to \infty} \mu_i(\{0\})$$

where μ is the spectral measure of $\cdot D$ associated with ε and μ_i is the spectral measure of $\cdot D_i$ associated with ε_i.

The rest of the proof goes through as before. We use the constant

$$d = k^2 \max_{r \in I_n} \{|H_r|\} \cdot \|D\|_1^2$$

and then Proposition 5.16 holds true because the finiteness of H_r implies that for all $g \notin H_r$ we can find N so large that $gh \notin G_i$ for all $i \geq N$ and all $h \in H_r$. Hence for each $g \notin H_r$ there is N such that $g \notin H_r G_i$ for all $i \geq N$. Proposition 5.18 holds true because we can consider the matrices D_i as symmetric $(\rho \times \rho)$-matrices with coefficients in $\mathbb{Z}$ where $\rho = \sum_{r \in I_n} [G : H_r G_i]$. Since $\rho \leq k[G : G_i]$, we obtain again the estimate

$$\frac{m_1 + \cdots + m_t}{[G : G_i]} \leq \frac{k \log d}{|\log \lambda|}. \qquad \qquad \square$$

Combining Theorem 5.19 with Theorem 4.14, we obtain the following version of Lück approximation for groups. As we already discussed in Sect. 4.4 of Chap. 4, the assumption of this theorem is often easier to establish in practice.

Theorem 5.20 *Let G be a residually finite group that has a finite type model for $\underline{E}G$. Then for any residual chain (G_i) and every $n \geq 0$, we have*

$$b_n^{(2)}(G) = \lim_{i \to \infty} \frac{b_n(G_i)}{[G : G_i]}.$$

In particular, Lück and Osin's groups from Sect. 5.4.1 above admit no model for *EG* with finite 2-skeleton either. It is a curious observation that all the properties of ℓ^2-Betti numbers gathered in Theorem 3.18 are immediate consequences of Theorem 5.19 in case G is residually finite (none of the properties were used in the proof).

5.4.3 Non-normal Subgroups

Consider a nested sequence

$$G = G_0 \geq G_1 \geq G_2 \geq \cdots$$

of not necessarily normal, finite index subgroups of G. Following [1, Section 3], we see that the chain (G_i) gives rise to a so-called *coset tree* T. Vertices of T are all right cosets $G_i g$. Two vertices $G_i g$ and $G_j h$ are connected by an edge if and only if $j = i + 1$ and $G_j h \subset G_i g$. The coset $G_0 = G$ provides a natural *root* for the tree so that the i-th level vertex set is just $G_i \backslash G$, and each node in $G_i \backslash G$ has precisely $[G_i : G_{i+1}]$ children. A typical coset tree is indicated in Fig. 5.2. We define the *boundary* ∂T of T as the set of all infinite rays in T starting at G_0. In other words, $\partial T = \varprojlim G_i \backslash G$ and this description makes sense not only in the category of sets but also in the category of topological spaces and of measure spaces. Each $G_i \backslash G$ carries the discrete topology and the uniform probability measure. Thus as a space, ∂T is compact, totally disconnected and Hausdorff. It has a basis of the topology given by *shadows* $\mathrm{sh}(G_i g)$ where a shadow $\mathrm{sh}(G_i g)$ consists of all rays going through the vertex $G_i g$. The Borel probability measure μ on ∂T is determined by the values $\mu(\mathrm{sh}(G_i g)) = \frac{1}{[G:G_i]}$. The group G permutes the cosets in $G_i \backslash G$ by right multiplication preserving the child–parent relation. Thus G acts on T from the right by tree automorphisms and we obtain an induced probability measure preserving right action of G on ∂T by homeomorphisms.

Lemma 5.21 *The action $\partial T \curvearrowleft G$ is* ergodic*: G-invariant measurable subsets of ∂T have measure 0 or 1.*

Fig. 5.2 The first three steps of a typical coset tree. Note that coset trees are generally not regular

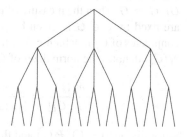

Proof If the chain stabilizes, the boundary is finite and the action is transitive so that the assertion is clear. Otherwise, Kuratowski's theorem [93, Theorem 15.6, p. 90] says ∂T is Borel-isomorphic to the unit interval with Lebesgue measure and so Lebesgue's density theorem applies: any measurable subset $A \subseteq \partial T$ is almost everywhere dense. In particular, if $\mu(A) > 0$, then for all $\varepsilon > 0$ there is some $G_i g_0$ with

$$\mu(\mathrm{sh}(G_i g_0) \cap A) > (1 - \varepsilon)\,\mathrm{sh}(G_i g_0).$$

If A is moreover G-invariant, then the same must hold for all $G_i g$. Adding up all these inequalities gives

$$\mu(A) = \sum_{G_i g \in G_i \backslash G} \mu(\mathrm{sh}(G_i g) \cap A) > 1 - \varepsilon,$$

which implies $\mu(A) = 1$.   □

Lemma 5.22 *If (G_i) is residual, then $\partial T \curvearrowleft G$ is free.*

Proof Given a nontrivial $g \in G$, there exists i such that $g \notin G_i$. Since G_i is normal, the element g permutes $G_i \backslash G$ without fixed points. Thus g moves all rays in ∂T.
                                                                                     □

The lemma leads naturally to the following weakening of a chain being residual.

Definition 5.23 A chain (G_i) of finite index subgroups of G is called *Farber* if the action $\partial T \curvearrowleft G$ is *essentially* free.

Of course "essentially free" means that μ-almost every point in ∂T has trivial stabilizer. To verify this condition, the following more explicit criterion is helpful. Let n_i be the number of subgroups conjugate to G_i in G. For $g \in G$, let $n_i(g)$ be the number of subgroups conjugate to G_i that contain g. For each $g \in G$, let $\mathrm{Fix}_{\partial T}(g)$ be the set of rays in ∂T fixed by g.

Proposition 5.24 *We have $\mu(\mathrm{Fix}_{\partial T}(g)) = \lim_{i \to \infty} \frac{n_i(g)}{n_i}$ and hence the chain (G_i) is Farber if and only if $\lim_{i \to \infty} \frac{n_i(g)}{n_i} = 0$ for all nontrivial $g \in G$.*

Proof Let $m_i(g)$ be the number of cosets fixed by g under the permutation action $G_i \backslash G \curvearrowleft G$. Then the measure of the set $P_i(g)$ of all paths in ∂T whose first i steps are fixed by $g \in G$ is given by $\mu(P_i(g)) = m_i(g)/[G : G_i]$. Each of the $n_i(g)$ conjugates of G_i in which g lies, fixes $[\mathcal{N}(G_i) : G_i]$ distinct cosets in $G_i \backslash G$, where $\mathcal{N}(G_i)$ denotes the normalizer of G_i in G. Hence

$$\frac{m_i(g)}{[G : G_i]} = \frac{n_i(g)[\mathcal{N}(G_i) : G_i]}{[G : \mathcal{N}(G_i)][\mathcal{N}(G_i) : G_i]} = \frac{n_i(g)}{n_i}. \tag{5.6}$$

Since $\mathrm{Fix}_{\partial T}(g) = \bigcap_i P_i(g)$ and the sets $P_i(g)$ are open and nested, the proposition follows from the outer regularity of μ.   □

Theorem 5.25 ([44]) *Let X be a free, finite type G-CW complex and let (G_i) be any Farber chain. Then for every $n \geq 0$ we have*

$$b_n^{(2)}(G \curvearrowright X) = \lim_{i \to \infty} \frac{b_n(G_i \backslash X)}{[G : G_i]}.$$

The original proof is given in [44, Theorem 0.3] but also a proof along the lines of Sect. 5.3 is possible. To establish weak convergence of spectral measures, one only has to observe that according to (5.6), the Farber condition says that the proportion of fixed points of the permutation that g defines on $G_i \backslash G$ becomes negligible for large i unless g is trivial. In other words, for all $D \in \mathbb{R}G$ the fraction

$$\frac{\operatorname{tr}_{\mathbb{R}}(\mathbb{R}(G_i \backslash G) \xrightarrow{\cdot D} \mathbb{R}(G_i \backslash G))}{[G : G_i]}$$

converges to the unit coefficient of D. It is also clear from (5.6) that each refinement of a Farber chain is again Farber. This shows that any Farber chain (G_i) can be turned into a residual chain by replacing each G_i with the *normal core* $\bigcap_{g \in G} g^{-1} G_i g$. Indeed, for normal subgroups the sequence $n_i(g)/n_i$ takes only the values 0 or 1, so that the Farber condition says the sequence eventually vanishes. Thus the total intersection of the normal cores of G_i is trivial. This means that Theorem 5.25 only applies to residually finite groups, just like Lück approximation does. Merely the permitted chains are more general.

In [17], Bergeron and Gaboriau settle the question in how far the Farber condition is optimal for approximating ℓ^2-Betti numbers. This includes the construction of examples of non-Farber chains which even violate Kazhdan's inequality. Nevertheless, it is shown that for every free, finite type G-CW complex X and every chain of finite index subgroups (G_i), the sequence $b_n(G_i \backslash X)/[G : G_i]$ converges. Generically, the limit will depend on the chain (G_i) and can be described in terms of X and ∂T.

5.4.4 Nontrivial Total Intersection

Given a chain (G_i) of finite index normal subgroups, it is apparent that the right hand side of Lück's approximation theorem is oblivious to proper coverings of $(\bigcap_i G_i) \backslash X$. Accepting that, we can formulate a version of the approximation theorem valid for all groups (which is however vacuous for Higman's group).

Theorem 5.26 *Let X be a free, finite type G-CW complex and let (G_i) be any chain of finite index normal subgroups. Set $K = \bigcap_i G_i$. Then for every $n \geq 0$, we have*

$$b_n^{(2)}(G/K \curvearrowright K \backslash X) = \lim_{i \to \infty} \frac{b_n(G_i \backslash X)}{[G : G_i]}.$$

Proof The proof of Proposition 5.16 in Sect. 5.3 needs the tiny modification that now gK being nontrivial in G/K says precisely that there is some i with $g \notin G_i$. The rest goes through as before. $\square$

5.4.5 Non-nested and Infinite Index Subgroups

This case is commonly subsumed under the term *approximation conjecture*. It has attracted quite some attention due to its intricate relation with the *determinant conjecture* dealing with "determinants" of matrices over the group ring. Moreover, the approximation conjecture gives some insight on the Atiyah conjecture 3.30 and whence on Kaplansky's Conjecture 1.1. These remarks call for a thorough discussion which we outsource to Sect. 5.6.

5.4.6 Further Variants

Lück's approximation theorem has become the prototype example of a whole multitude of results recognizing ℓ^2-invariants as limits of finite dimensional counterparts. We only mention a few here and come back to this aspect in Chap. 6 when we discuss the asymptotics of torsion in homology.

In the default setting of a free, finite type G-CW complex X and a residual chain (G_i), we can consider any field k and set $b_n(G_i \backslash X; k) = \dim_k H_n(G_i \backslash X; k)$. Of course, this only gives something new if k has positive characteristic p and then $b_n(G_i \backslash X; k) = b_n(G_i \backslash X; \mathbb{F}_p)$ where $\mathbb{F}_p$ is the field with p elements. In the case of positive characteristic, convergence of the sequence $b_n(G_i \backslash X; k)/[G : G_i]$, let alone independence of (G_i), is wide open for general residually finite G. But in the special case when G is torsion-free and elementary amenable (see p. 60), Linnell et al. [107] show

$$\dim_{kG}^{\mathrm{Ore}} H_n(X; k) = \lim_{i \to \infty} \frac{b_n(G_i \backslash X; k)}{[G : G_i]}$$

for any k. Here the left hand side denotes the *Ore dimension* of the kG-module $H_n(X; k)$. If G is torsion-free elementary amenable, the group ring kG, though possibly noncommutative, can be localized at $S = kG \setminus \{0\}$ to a skew field $S^{-1}kG$ and then

$$\dim_{kG}^{\mathrm{Ore}} H_n(X, k) = \dim_{S^{-1}kG}(S^{-1}kG \otimes_{kG} H_n(X; k)).$$

Observe that the Ore dimension is by definition always an integer. In the characteristic zero case, this is in accordance with Linnell's Theorem 3.33 which says in particular that the Atiyah conjecture 3.30 with $R = \mathbb{Q}$ holds for torsion-free elementary amenable groups.

Every elementary amenable group is *amenable*. It admits a left G-invariant bounded linear functional μ on the Banach space $\ell^\infty G$ with $\mu(1) = 1$. Grigorschuk's group [61] of intermediate growth is an amenable group which is not elementary amenable. For amenable groups, Dodziuk–Mathai gave an approximation theorem for ℓ^2-Betti numbers in terms of subcomplexes of X, rather than quotients. The interested reader may find out about this in [36].

The growth of Betti numbers has also been examined in more specific geometric situations. For example, if G is a discrete, cocompact subgroup of $\mathrm{SL}(k, \mathbb{R})$ with $k \geq 3$, then G acts by isometries on the contractible *symmetric space* $X = \mathrm{SL}(k, \mathbb{R})/\mathrm{SO}(k)$. By discreteness, G intersects the compact group $\mathrm{SO}(k)$ in a finite group. This implies that X is proper and in fact, after choosing a suitable G-CW structure, a finite model for $\underline{E}G$. It follows from Borel [21] that $b_n^{(2)}(G) = 0$ for $n \geq 0$.

Theorem 5.27 (Abért et al. [2]) *For G as above, let (G_i) be any sequence of distinct, finite index subgroups of G (not necessarily nested, not necessarily normal). Then for every $n \geq 0$, we have*

$$\lim_{i \to \infty} \frac{b_n(G_i)}{[G : G_i]} = 0.$$

The background to this astonishing result is that in the situation at hand, the condition $[G : G_i] \to \infty$ is enough to ensure that the coverings $G_i \backslash X$ converge to X in the sense of *Benjamini–Schramm*: for every $R > 0$, we have

$$\lim_{i \to \infty} \frac{\mathrm{vol}((\Gamma_i \backslash X)_{<R})}{\mathrm{vol}(\Gamma_i \backslash X)} = 0$$

where $(\Gamma_i \backslash X)_{<R}$ denotes the *R-thin part* consisting of the points in $\Gamma_i \backslash X$ with *injectivity radius* $< R$ (the maximal radius for which the exponential map is a diffeomorphism). This and a variety of other highly interesting and much related theorems can be found in the influential paper [2].

To conclude this section, let me report on two most recent approximation results of Kionke. The first one [97] is concerned with approximating *multiplicities* of finite group representations. Let H be a finite group and let X be a finite H-CW complex. Then the homology $H_n(X; \mathbb{C})$ is a finite dimensional representation of H which therefore decomposes as a direct sum of irreducibles χ with multiplicities $m(\chi, H_n(X; \mathbb{C}))$. Kionke defines an ℓ^2-counterpart $m^{(2)}(\chi, X; G)$ of these multiplicities for a proper, finite type G-CW complex X and shows that for any residual chain (G_i) we have

$$\lim_{i \to \infty} \frac{m(\chi, H_n(G_i \backslash X; \mathbb{C}))}{[G : G_i]} = m_n^{(2)}(\chi, X; G).$$

The starting point for the second result is the observation that in Lück's approximation theorem, the real number $b_n^{(2)}(X)$ is the limit of the sequence of rational numbers $b_n(G_i \backslash X)/[G : G_i]$ when considering $\mathbb{Q}$ as a subspace of $\mathbb{R}$. Number theory philosophy says however, that the p-adic numbers $\mathbb{Q}_p$ are completions of $\mathbb{Q}$ with equal rights. It turns out that the sequence of Betti numbers $b_n(G_i \backslash X)$ converges in $\mathbb{Q}_p$ if one does *not* divide by the index. For more on this interesting idea, the reader is referred to [94].

5.5   Rank Gradient and Cost

Related to the notion of deficiency studied in Sect. 4.5.2 is the *rank* $d(G)$ of a group defined as the minimal cardinality of a generating subset of G. If G is finitely generated and $P = \langle S|R \rangle$ is a presentation with $|S| = d(G)$, we can again form the presentation complex X_P which now is a possibly infinite two-dimensional CW complex with one 0-cell and $d(G)$ 1-cells. Corresponding to any finite index subgroup $H \leq G$, we have a finite sheeted covering X_H of X_P whose lifted CW structure has $[G : H]$ many 0-cells and $d(G) \cdot [G : H]$ many 1-cells. Hence the 1-skeleton $(X_H)_1$, being a connected graph with $[G : H]$ vertices, possesses a *spanning tree* (a contractible subgraph containing all vertices) with $[G : H] - 1$ edges. Since subcomplexes are cofibrations, the homotopy type of X_H remains unchanged when collapsing the spanning tree, so the fundamental group of the resulting space is still isomorphic to H. This shows the inequality

$$d(H) \leq d(G)[G : H] - ([G : H] - 1) = (d(G) - 1)[G : H] + 1. \qquad (5.7)$$

Lackenby's *rank gradient* quantifies how far away from equality the inequality can get [100].

Definition 5.28 The *(absolute) rank gradient* of the group G is given by

$$\mathrm{RG}(G) = \inf_{\substack{H \leq G \\ [G:H] < \infty}} \frac{d(H) - 1}{[G : H]}.$$

The extreme case $\mathrm{RG}(G) = 0$ occurs for example for *mapping torus groups* of the form $G = K \rtimes \mathbb{Z}$ for any finitely generated group K. We can find a self-homotopy equivalence f of a CW model BK with one 0-cell and $d(K)$ many 1-cells such that $\pi_1 f$ is the automorphism defining the semidirect product G. As part of Exercise 4.5.1, or knowing about the long exact sequence of homotopy groups for a fibration, we see that the mapping torus $T(f)$ is a model for BG. From the proof of Theorem 3.38, we thus extract that $T(f^k)$ is a model for $B(K \rtimes (k\mathbb{Z}))$ with $d(K)+1$ many 1-cells, independently of k. Hence the index k subgroups $K \rtimes (k\mathbb{Z}) \leq G$ reveal that $\mathrm{RG}(G) = 0$.

Another class of groups with vanishing rank gradient is formed by so called *S-arithmetic groups* with trivial *congruence kernel*. As above, vanishing of the rank gradient follows from a uniform upper bound on the rank of a certain sequence of subgroups: the *principal congruence subgroups*. The relevant definitions will be given in Sects. 6.6 and 6.7. The bound on the rank was found by Sury and Venkataramana [164] in the example of $SL_n(\mathbb{Z})$ for $n \geq 3$, and stated in general. Moreover, *Artin groups* with connected defining graph, $Aut(F_n)$ for $n \geq 2$, $Out(F_n)$ for $n \geq 3$, and *mapping class groups* $MCG(\Sigma_g)$ of surfaces with genus $g \geq 2$ all have zero rank gradient [92].

On the other hand, for the free group F_k on k letters, we have $RG(F_k) = k - 1$ because in that case, inequality (5.7) is an equality. We only have to notice that the first Betti number provides a lower bound for the rank of a group. Hence for a finite index subgroup $H \leq F_k$, we have

$$d(H) \geq b_1(H) = 1 - \chi(H) = 1 - [F_k : H](1 - k) = (k - 1)[F_k : H] + 1.$$

Without much trouble, one concludes from this that the index 12 overgroup $SL_2(\mathbb{Z})$ of F_2 likewise has positive rank gradient [100, Lemma 3.1]. More generally than nonabelian free groups, finitely presented groups with $def(G) \geq 2$ have positive rank gradient. To see this, let $P = \langle S | R \rangle$ be a presentation realizing the deficiency. Then the covering X_H of X_P constructed as above has $[G : H] \cdot |R|$ many 2-cells, hence

$$b_1(H) \geq ((|S| - 1)[G : H] + 1) - [G : H]|R| = [G : H](|S| - |R| - 1) + 1$$

which gives $RG(G) \geq def(G) - 1$. Comparing this inequality with Theorem 4.20 and reviewing the examples given so far, one realizes that $RG(G)$ bears quite some resemblance to the first ℓ^2-Betti number $b_1^{(2)}(G)$. Indeed, Lück's approximation theorem has the following consequence.

Theorem 5.29 *Every finitely presented residually finite group G satisfies*

$$RG(G) \geq b_1^{(2)}(G).$$

Proof Inequality (5.7) implies that the net $\left(\frac{d(H)-1}{[G:H]}\right)$, indexed by all finite index subgroups $H \leq G$ and directed by containment, is monotone decreasing. As it is bounded from below by zero, it converges and the limit is $RG(G)$. If we define $H_i \leq G$ as the intersection of the (finitely many!) normal subgroups of G of index at most i, then $(H_i)_{i=1}^{\infty}$ is a residual chain such that the sequence $\left(\frac{d(H_i)-1}{[G:H_i]}\right)$ is a cofinal subnet of $\left(\frac{d(H)-1}{[G:H]}\right)$ by the normal core construction. Hence $RG(G)$ equals the limit of this subnet. Since G is finitely presented, it has a model for EG with finite 2-skeleton so that the inequality $d(H_i) \geq b_1(H_i)$ and Theorem 5.3 complete the proof: We only have to recall from the beginning of Sect. 5.3 that the conclusion of Theorem 5.3 on the n-th ℓ^2-Betti number holds if the $(n + 1)$-skeleton of EG is finite. □

It comes as a little surprise that an intriguing group invariant from measurable dynamics, called the *cost* of G, can be squeezed into the inequality of Theorem 5.29:

$$RG(G) \geq \text{cost}(G) - 1 \geq b_1^{(2)}(G) \tag{5.8}$$

for every infinite, finitely generated, residually finite group G, even if we drop the assumption of finite presentation. Cost of a group was introduced by Gaboriau [54] building on the earlier notion for equivalence relations by Levitt [101]. To prepare the definition, let (Ω, μ) be a standard Borel probability space which again by Kuratowski's theorem is Borel isomorphic to the Lebesgue probability space $([0, 1], \lambda)$. Say (Ω, μ) comes with an ergodic probability measure preserving (p.m.p.) right action $\Omega \curvearrowleft G$ by a countably infinite discrete group G. Then "being in the same G-orbit" defines the measurable *orbit equivalence relation* $E \subset \Omega \times \Omega$. We interpret any measurable subset $S \subset \Omega \times \Omega$ as the set of edges defining a directed graph with vertex set Ω. In this way, our equivalence relation E defines an uncountable directed graph whose connected components are countable *complete digraphs*: Any two distinct vertices in a connected component are joined by two unique edges of opposite direction. If $S \subset E$ is a measurable subset, we call S a *subgraph* of E. Given a graph $S \subset \Omega \times \Omega$ and $k \geq 1$, we define the k-th *power* of S by agreeing that $(x, y) \in S^k$ if and only if either $x = y$ or there exists an undirected path in S from x to y of length at most k. If we have $E = \bigcup_{k \geq 1} S^k$, we write $E = \langle S \rangle$ and say that $S \subset E$ *spans* E.

Definition 5.30 The *edge measure* $e(S)$ of a subgraph $S \subset E$ is given by

$$e(S) = \int_\Omega \deg_S(x) \, d\mu(x)$$

where $\deg_S(x) = |\{y \in \Omega : (x, y) \in S\}|$ is the number of outgoing edges from x.

So the edge measure $e(S)$ is the average number of outgoing edges per vertex in the graph S. As such, it is a measure of complexity for S. This means we can consider the infimum of the edge measures over all spanning subgraphs of E as the price we need to pay for generating the equivalence relation of the measurable group action.

Definition 5.31 The *cost* of $\Omega \curvearrowleft G$ is the *cost* of the measurable equivalence relation $E \subset \Omega \times \Omega$ and it is given by $\text{cost}(E) = \text{cost}(\Omega \curvearrowleft G) = \inf_{\langle S \rangle = E} e(S)$. We set

$$\text{cost}(G) = \inf \text{cost}(\Omega \curvearrowleft G)$$

where the infimum is taken over all ergodic and essentially free p.m.p. actions $\Omega \curvearrowleft G$.

Encountering the definition for the first time, one could get the idea to pick $S \subset E$ consisting of an oriented path in each connected component of E that would travel precisely once through each vertex of the complete graph to conclude that cost is always one. But that is of course nonsense because the dependence of cost on the group action is encoded in the measurability condition on the subset $S \subset E$. The above "picking" of S is a perfect application of the axiom of choice from which one cannot expect any measurability assertion on S whatsoever. If however G is finitely generated by $g_1, \ldots, g_n$, then the edges (x, xg_i) for $x \in \Omega$ and $i = 1, \ldots, n$ form a measurable spanning subgraph $S \subset E$, implying that $\text{cost}(G) \leq d(G)$. The connected components of almost all points in S will look like the Cayley graph of G without distinguished base point. If for example $G \cong F_n$ is free on $g_1, \ldots, g_n$, this means S is essentially a *forest*: The connected component of a vertex $x \in \Omega$ is almost surely a tree. In his fundamental work on cost [54], D. Gaboriau has shown that this subgraph S realizes the cost so that $\text{cost}(\Omega \curvearrowleft F_n) = n$ for all essentially free ergodic p.m.p. actions $\Omega \curvearrowleft F_n$ and in particular $\text{cost}(F_n) = n$. It is unknown whether the phenomenon of constant cost for all these actions is observable for all groups:

Question 5.32 (Fixed Price Problem) Does every essentially free ergodic probability measure preserving action $\Omega \curvearrowleft G$ of a countable group G have the same cost?

It is a famous theorem of Ornstein and Weiss [140, Theorem 6] that the orbit equivalence relation of a p.m.p. action $\Omega \curvearrowleft G$ of an amenable group G with almost surely infinite orbits can be generated by a single Borel automorphism and hence by an action $\Omega \curvearrowleft \mathbb{Z}$. Since $\mathbb{Z}$ is a free group, this demonstrates that infinite amenable groups have fixed price one.

We will next present a beautiful theorem due to Abért and Nikolov [1] which builds the bridge from rank gradient to cost, hence from a combinatorial to a dynamic invariant. To state the theorem, we define the *relative rank gradient* of G with respect to any chain (G_i) of finite index subgroups by

$$\text{RG}(G, (G_i)) = \lim_{i \to \infty} \frac{d(G_i) - 1}{[G : G_i]}.$$

We now assume for the rest of this section that G is finitely generated. This has the effect that the canonical residual chain (H_i) in G from the proof of Theorem 5.29 is defined and we have $\text{RG}(G, (H_i)) = \text{RG}(G)$. Recall from Sect. 5.4.3 that any chain (G_i) defines the ergodic p.m.p. action $\partial T \curvearrowleft G$ on the boundary ∂T of the coset tree T. For the canonical chain (H_i), the boundary of the coset tree ∂T coincides with the *profinite completion* $\widehat{G}$ of G to be examined more closely in Sect. 6.7. Assuming that G is also residually finite, we can choose a Farber sequence (G_i) of finite index subgroups in G so that the action $\partial T \curvearrowleft G$ is essentially free.

Theorem 5.33 (Abért–Nikolov) *Let (G_i) be Farber in G with coset tree T. Then*

$$\text{cost}(\partial T \curvearrowleft G) = \text{RG}(G, (G_i)) + 1.$$

We obtain the first inequality in (5.8) as an immediate corollary and this inequality is actually an equality if the fixed price problem has an affirmative answer. As another consequence of the theorem, the fixed price problem would positively answer the open question whether the relative rank gradient is in fact independent of the Farber chain. On the other hand, Abért–Nikolov explain that the hyperbolic *rank vs Heegard genus conjecture* would imply that the relative rank gradient does depend on the Farber chain; so either the rank vs Heegard conjecture or the fixed price problem is false. Every closed orientable 3-manifold M admits a decomposition into two handlebodies along some closed surface of genus g. The minimal possible such g is called the *Heegard genus* $g(M)$. The core of any one handlebody provides a set of g generators for $\pi_1 M$ so that $g(M) \geq d(\pi_1 M)$. The rank vs Heegard conjecture asks whether equality holds. Amusingly, the case $d(\pi_1 M) = 0$ gives the Poincaré conjecture. After several non-hyperbolic counterexamples had been constructed, Li [102] has meanwhile also disproved the hyperbolic case so that the fixed price problem is the one that remains open.

The proof of Theorem 5.33 we are about to give has previously appeared as the blog post [86]. We will see in a moment that the following notion closely related to measurable (sub-)graphs shall come in handy: A *graphing* is a measurable subset $M \subset \Omega \times G$. It suggests itself to picture an element $(x, g) \in \Omega \times G$ as an "arrow" in Ω pointing from x to xg. Note that (almost all) these arrows are determined by their initial and final point if and only if the action $\Omega \curvearrowleft G$ is (essentially) free. We can sort the arrows in the subset M either by initial point or by direction: either by the Ω- or by the G-coordinate. So interchangeably we think of M as a family of subsets

$$M_g = \{x \in \Omega : (x, g) \in M\} \subset \Omega$$

parametrized by group elements $g \in G$, or as a family of subsets

$$M_x = \{g \in G : (x, g) \in M\} \subset G$$

parametrized by points $x \in \Omega$. Guided by what we did above, we define the *k-th power* M^k of M by all the arrows we obtain by composing up to k arrows from M regardless of their direction. In more mathematical terms this means $(x, g) \in M^k$ if and only if there is $0 \leq l \leq k$ and a decomposition $g = g_1 \cdots g_l$ in G such that for all $0 \leq i < l$ either

$$(xg_1 \cdots g_i, g_{i+1}) \in M \quad \text{or} \quad (xg_1 \cdots g_{i+1}, g_{i+1}^{-1}) \in M. \tag{5.9}$$

Note that $M^0 = \Omega \times \{1\}$, regardless of what M is. We say that a graphing $M \subset \Omega \times G$ *spans* $\Omega \times G$ if we have $\Omega \times G = \bigcup_{k \geq 0} M^k$. In this case we write $\langle M \rangle = \Omega \times G$. Let e be the measure on $\Omega \times G$ given by the product of μ and the counting measure on G.

Definition 5.34 The *groupoid cost* of the ergodic p.m.p. action $\Omega \curvearrowright G$ is

$$\mathrm{gcost}(\Omega \curvearrowright G) = \inf_{\langle M \rangle = \Omega \times G} e(M).$$

To explain the terminology, we observe that the set $\Omega \times G$ has a groupoid structure: two arrows $(x_1, g_1), (x_2, g_2) \in \Omega \times G$ can be composed if and only if $x_1 g_1 = x_2$, meaning the first arrow points to the initial point of the second, and in that case their composition is $(x_1, g_1 g_2)$. So a graphing spans $\Omega \times G$ if and only if it generates $\Omega \times G$ as groupoid.

Proposition 5.35 *We have*

$$\mathrm{gcost}(\Omega \curvearrowright G) \geq \mathrm{cost}(\Omega \curvearrowright G)$$

with equality if the action is essentially free.

Proof A graphing $M \subset \Omega \times G$ defines a subgraph $\Phi(M) \subset E$ of the orbit equivalence relation $E \subset \Omega \times \Omega$ by setting

$$\Phi(M) = \{(x, xg) \colon (x, g) \in M\}.$$

Clearly $\Phi(M^k) = \Phi(M)^k$ so that Φ preserves the spanning property. We have $\deg_{\Phi(M)}(x) \leq |M_x|$ with equality almost everywhere if the action $\Omega \curvearrowright G$ is essentially free. Integrating over Ω gives the inequality. To obtain equality for essentially free actions one still has to show that each spanning subgraph can be obtained from a spanning graphing via Φ; we skip this argument which needs some technical care but no unusual ideas [1, Lemma 6]. □

Theorem 5.36 *If (G_i) is a chain of finite index subgroups in G with coset tree T, then*

$$\mathrm{gcost}(\partial T \curvearrowright G) = \mathrm{RG}(G, (G_i)) + 1.$$

We stress that for this result, the subgroups G_i must not be normal, the chain must not be Farber, and the subgroups are not required to have trivial total intersection. Clearly, this theorem and Proposition 5.35 complete the proof of Theorem 5.33.

Proof We first show $\mathrm{gcost}(\partial T \curvearrowright G) \leq \mathrm{RG}(G, (G_i)) + 1$. For all $\varepsilon > 0$ we find some i with $\frac{d(G_i) - 1}{[G : G_i]} < \mathrm{RG}(G, (G_i)) + \varepsilon$. Thus the integer

$$d = \lfloor (\mathrm{RG}(G, (G_i)) + \varepsilon)[G : G_i] \rfloor + 1$$

gives an upper bound for $d(G_i)$. Say G_i is generated by $g_1, \ldots, g_d$ and let $1 = \gamma_1, \ldots, \gamma_{[G : G_i]}$ be a system of representatives for $G_i \backslash G$. We define a graphing

$M \subseteq \partial T \times G$ by setting

$$M_g = \begin{cases} \mathrm{sh}(G_i) & \text{if } g = g_j \text{ for some } j \geq 1 \text{ or } g = \gamma_j \text{ for some } j > 1, \\ \emptyset & \text{otherwise.} \end{cases}$$

We claim that $\langle M \rangle = \partial T \times G$. Indeed, let $(x, g) \in \partial T \times G$ and let γ_a and γ_b be the representatives from the list for which $x \in \mathrm{sh}(G_i \gamma_a)$ and $\gamma_a g \gamma_b^{-1} \in G_i$. Hence we can write g as a word of the form $\gamma_a^{-1} g_{j_1}^{\pm 1} \cdots g_{j_k}^{\pm 1} \gamma_b$. With respect to this factorization of g one easily verifies the criterion (5.9) to conclude $(x, g) \in M^{k+2}$ proving the claim. By definition M_g equals $\mathrm{sh}(G_i)$ for precisely $d + [G : G_i] - 1$ elements in G and is empty otherwise. Hence the graphing M has measure

$$e(M) = \frac{d + [G : G_i] - 1}{[G : G_i]}.$$

It follows that

$$\mathrm{gcost}(\partial T \curvearrowright G) \leq \frac{d - 1}{[G : G_i]} + 1 \leq \mathrm{RG}(G, (G_i)) + 1 + \varepsilon.$$

The reverse inequality $\mathrm{gcost}(\partial T \curvearrowright G) \geq \mathrm{RG}(G, (G_i)) + 1$ is somewhat harder. Given $\varepsilon > 0$, there exists a graphing M which spans $\partial T \times G$ and has measure $e(M) < \mathrm{gcost}(\partial T \curvearrowright G) + \frac{\varepsilon}{2}$. The first thing to do now is to construct yet another graphing $N \subseteq \partial T \times G$ which is close to M in the sense that for the symmetric difference we have $e(N \triangle M) < \frac{\varepsilon}{2}$ and such that N has the convenient property that each N_g is a finite union of shadows which is nonempty only for finitely many $g \in G$. Since the shadows form a countable basis of the topology of ∂T, it is conceivable that such a "finite approximation" to M exists. So we shall allow ourselves to skip the precise technical construction [1, Lemma 5]. Since N is made up from only finitely many shadows altogether, there exists a large enough i such that each N_g is in fact a finite union of level-i shadows of the form $\mathrm{sh}(G_i h)$.

We define a finite, directed, labeled graph $\mathcal{G}$ as follows. The vertex set is $V = G_i \backslash G$ and for each $g \in G$ we connect $w \in V$ with $wg \in V$ by an edge of label g if and only if $\mathrm{sh}(w) \subseteq N_g$. The graph $\mathcal{G}$ has the canonical base point $v = G_i \in V$. This data defines a homomorphism of groups

$$\varphi \colon \pi_1(\mathcal{G}, v) \longrightarrow G$$

$$l = (e_1, \ldots, e_k) \longmapsto \mathrm{label}(e_1)^{\pm 1} \cdots \mathrm{label}(e_k)^{\pm 1}$$

which multiplies the labels along a loop of edges, inverting the label whenever we travel through an edge in reverse direction. We claim that the image of the homomorphism φ is precisely G_i. Indeed, for each $l \in \pi_1(\mathcal{G}, v)$ we have $v\varphi(l) = v$ by the construction of the graph $\mathcal{G}$. Thus $\varphi(l) \in \mathrm{Stab}(v) = G_i$. Let $h \in G_i$ be any element and pick some ray $x \in \mathrm{sh}(v)$. Since N spans $\partial T \times G$, there is a factorization

$h = g_1 \cdots g_k$ with $g_j \in G$ such that for all $0 \leq j < k$ either

$$(xg_1 \cdots g_j, g_{j+1}) \in N \quad \text{or} \quad (xg_1 \cdots g_{j+1}, g_{j+1}^{-1}) \in N.$$

For $0 \leq j \leq k$, let $w_j \in V = G_i \backslash G$ be the level-i vertex in the coset tree T through which the ray $x\gamma_1 \cdots \gamma_j$ passes. Then we have $w_0 = w_k = v$ and for each $0 \leq j < k$ either w_j is connected to w_{j+1} by an edge in $\mathcal{G}$ with label g_{j+1} or w_{j+1} is connected to w_j by an edge of label g_{j+1}^{-1}. Hence these edges form a loop l with $\varphi(l) = h$ proving the claim.

Thus G_i is a quotient group of $\pi_1(\mathcal{G}, v)$. The latter group is free of rank $1 - \chi(\mathcal{G})$. Note that

$$e(N) = \sum_{g \in G} \sum_{\text{sh}(G_i h) \subseteq N_g} \frac{1}{[G : G_i]}$$

so that $e(N)[G : G_i]$ is the number of edges in $\mathcal{G}$ while the number of vertices in $\mathcal{G}$ is of course $[G : G_i]$. Putting pieces together, we obtain

$$d(G_i) \leq d(\pi_1(\mathcal{G}, v)) = 1 + e(N)[G : G_i] - [G : G_i].$$

By subadditivity of the measure e applied to $N \subseteq M \cup (N \triangle M)$ we conclude

$$\text{RG}(G, (G_i)) \leq \frac{d(G_i) - 1}{[G : G_i]} = e(N) - 1 < \text{gcost}(\partial T \curvearrowleft G) + \varepsilon - 1$$

and the proof is complete. □

The second inequality of (5.8), or better the inequality

$$\text{cost}(G) - 1 \geq b_1^{(2)}(G) - b_0^{(2)}(G)$$

to include finite groups into the discussion, is due to Gaboriau [55, Corollaire 3.23 and Corollaire 3.16]. It holds for all (at most) countable discrete groups G. Hence the above mentioned Theorem of Ornstein–Weiss implies that all amenable groups G have $b_1^{(2)}(G) = 0$ and in fact [55, Corollaire 0.1] that all amenable groups are ℓ^2-acyclic (in the sense of p. 77). This result was first proven by Cheeger and Gromov [29, Theorem 0.2]. By a spectral sequence argument, one can conclude from this more generally that any group G with an infinite amenable normal subgroup is ℓ^2-acyclic [117, Lemma 6.66, p. 271]. This applies for example to all groups with infinite center. It was moreover long known that groups with Kazhdan's property (T) [14], a property often described as the "opposite" of being amenable, have vanishing first ℓ^2-Betti number [13, Corollary 6] while they may or may not have some positive higher ℓ^2-Betti number. The corresponding statement that (T)-groups have cost one has only recently been verified [77] but the fixed price problem remains open for them. It is yet unknown whether either inequality in (5.8) can be strict.

There is a lot more to say about ℓ^2-invariants and measured group theory. We invite the reader to learn about it from the inspiring survey articles of Furman [53] and Gaboriau [56, 57]. As evidence of the potential of measured group theory methods, let us conclude this section with a geometric result [155, Corollary to Theorem A] that Sauer obtained by applying Gaboriau's theory of $\mathscr{R}$-simplicial complexes [55, Section 2] and other tools from measurable group theory to elaborate a proof strategy anticipated by Gromov [66, p. 297] upon ideas of A. Connes. Given a smooth manifold M, the *minimal volume* $\min\mathrm{vol}(M)$ is the greatest lower bound of the volumes of all complete Riemannian metrics on M with sectional curvature pinched between -1 and 1.

Theorem 5.37 (Main Inequality for ℓ^2-Betti Numbers) *For each dimension d, there is a constant $C_d > 0$ such that every closed aspherical smooth d-manifolds M satisfies*

$$b_n^{(2)}(\tilde{M}) \leq C_d \min\mathrm{vol}(M) \quad \text{for all } n \geq 0.$$

As M is aspherical, we have $b_n^{(2)}(\tilde{M}) = b_n^{(2)}(\pi_1 M)$. So the theorem remarkably identifies an *orbit equivalence invariant* of $\pi_1 M$ [55, Corollaire 3.16] as a lower bound on the minimal volume of M, the latter being an intrinsically geometric concept.

5.6 Approximation, Determinant, and Atiyah Conjecture

The formulation of Lück's approximation theorem given in (5.2) still makes sense if the normal subgroups G_i of G have infinite index. Just notice that ℓ^2-Betti numbers of the G/G_i-CW complex $G_i\backslash X$ are defined regardless of whether G_i has finite or infinite index. One might also come up with the idea to not only consider limits of sequences on the right hand side but also limits of nets (see p. 16) over *residual systems* $(G_i)_{i\in I}$ of normal subgroups directed by containment "$\supseteq$" with $\bigcap_{i\in I} G_i = \{1\}$. The corresponding approximation statement has become known as the *approximation conjecture*.

Conjecture 5.38 (Approximation Conjecture) Let X be a free, finite type G-CW complex and let $(G_i)_{i\in I}$ be a residual system. Then for every $n \geq 0$ we have

$$b_n^{(2)}(G \curvearrowright X) = \lim_{i \in I} b_n^{(2)}(G/G_i \curvearrowright G_i\backslash X).$$

Similar to Lück's approximation theorem and to the Atiyah conjecture, the approximation conjecture is in fact not so much a topological question but more an algebraic one. In fact, the following version is equivalent as a consequence of Proposition 3.29.

Conjecture 5.39 Let G be a group with residual system $(G_i)_{i \in I}$. Then for all $A \in M(k, l; \mathbb{Q}G)$ with reductions $A_i \in M(k, l; \mathbb{Q}(G/G_i))$, we have

$$\dim_{\mathcal{R}G} \ker(\ell^2 G^k \xrightarrow{\cdot A} \ell^2 G^l) = \lim_{i \in I} \dim_{\mathcal{R}(G/G_i)} \ker(\ell^2(G/G_i)^k \xrightarrow{\cdot A_i} \ell^2(G/G_i)^l).$$

Considering coefficients in $\mathbb{Q}$ instead of $\mathbb{Z}$ is possible because scalar multiplication with the l.c.m. of the denominators does not alter the kernels. Note that Proposition 3.29 requires G to be finitely generated but since A has only finitely many entries, it lies in $M(k, l; \mathbb{Q}H)$ for a finitely generated subgroup H of G and the von Neumann dimension of $\ker(\cdot A)$ is the same over $\mathcal{R}(G)$ and $\mathcal{R}(H)$ as we saw as part of Exercise 3.3.3.

Allowing infinite index normal subgroups has a remarkable advantage: the quotient groups G/G_i can be torsion-free and this permits the following application of the approximation conjecture to the Atiyah conjecture 3.30.

Theorem 5.40 *Let G be a group with residual system (G_i) satisfying the approximation conjecture. If each group G/G_i is torsion-free and satisfies the Atiyah conjecture 3.30 with $R = \mathbb{Q}$, then the same is true for G.*

Proof Any torsion element $g \in G$ becomes trivial in all quotient groups G/G_i as these are torsion-free. Hence $g \in \bigcap_i G_i$ is trivial and G is torsion-free. By assumption, for any $A \in M(k, l; \mathbb{Q}G)$, the net $(\dim_{\mathcal{R}(G/G_i)}(\ker \cdot A_i))_{i \in I}$ consists of integers. Hence if G satisfies the approximation conjecture, then $\lim_{i \in I} \dim_{\mathcal{R}(G/G_i)}(\ker \cdot A_i) = \dim_{\mathcal{R}(G)}(\ker \cdot A)$ is an integer as well. $\square$

This approach to the Atiyah conjecture is due to Schick [156, 158]. In view of Theorem 3.32, it should be enough motivation to tackle the approximation conjecture. Revisiting Sect. 5.3, we see that the framework of the proof of Lück's approximation theorem remains valid verbatim for the approximation conjecture if we only replace limits of sequences with limits of nets. Also the proof of the Portmanteau theorem works equally well for nets. Hence, showing the approximation conjecture amounts to establishing Propositions 5.16 and 5.18 in the new situation. For the first proposition, which asserts weak convergence of spectral measures, this is trouble-free: $\bigcap_{i \in I} G_i$ is trivial, hence traces converge. The crux of the matter is the second proposition. Recall that it sets up a uniform logarithmic bound for spectral distribution functions from an innocuous observation in (5.5): the product of positive eigenvalues of D_i is an integer, hence uniformly bounded from below by one. This argument breaks down when the quotient groups G/G_i are infinite. So the first step for proving the approximation conjecture consists in finding a reformulation of (5.5) that would still make sense when the matrices A_i (or better $A_i^* A_i$) act on infinite dimensional Hilbert spaces. To this end, we notice that the product $\lambda_1^{m_1} \cdots \lambda_s^{m_s}$ in (5.5) is precisely the *determinant* of the operator $\cdot D_i$ when we restrict domain and target to the orthogonal complement of the kernel of $\cdot D_i$.

So let us try and define such a "determinant" in the general setting of a morphism $T : H \to K$ of finitely generated Hilbert $\mathcal{L}(G)$-modules H and K. The operator

$|T|$ was already constructed in Exercise 2.2.5. Alternatively, we apply continuous functional calculus (Theorem 5.6) to the self-adjoint operator T^*T on H and obtain $|T| = \sqrt{T^*T}$. The operator $|T|$ is then positive by Proposition 5.7. It is moreover G-equivariant as it lies in the von Neumann algebra generated by the G-equivariant operator T^*T. Hence also the measure $P_{|T|}$ from p. 95 takes values in G-equivariant projections so that we can take the von Neumann trace to obtain a canonical real valued measure $\mu_{|T|} = \operatorname{tr}_{\mathcal{R}(G)} P_{|T|}$ on $\sigma(|T|)$. By construction, it is the spectral measure $\mu_{|T|} = \mu_{\varepsilon,|T|}$ where $\varepsilon \in H$ is the preimage of $\operatorname{pr}(e \oplus \cdots \oplus e) \in (\ell^2 G)^k$ under any embedding $i : H \hookrightarrow (\ell^2 G)^k$ where "pr" is the orthogonal projection onto $i(H)$. Indeed, the definition of $\operatorname{tr}_{\mathcal{R}(G)}$ in Proposition 2.33 shows that

$$\operatorname{tr}_{\mathcal{R}(G)} P_{|T|} = \langle e \oplus \cdots \oplus e, (i \circ P_{|T|} \circ \operatorname{pr})(e \oplus \cdots \oplus e)\rangle =$$
$$= \langle \operatorname{pr}(e \oplus \cdots \oplus e), P_{|T|}(\operatorname{pr}(e \oplus \cdots \oplus e))\rangle$$

because $i^* = \operatorname{pr}$. Note that the notation $\mu_{|T|}$ intentionally collides with our earlier notation for a basic measure of $|T|$ from p. 93.

Proposition 5.41 *The spectral measure $\mu_{|T|} = \mu_{\varepsilon,|T|}$ is basic for $|T|$.*

Proof Let $x \in H$ be a nonzero vector and let $A \subset \sigma(|T|)$ be measurable with $\mu_{x,|T|}(A) > 0$. Since $P_{|T|}(A) = P_{|T|}(A)^2 = P_{|T|}(A)^*$ is an orthogonal projection, we obtain

$$0 < \mu_{x,|T|}(A) = \langle x, P_{|T|}(A)x\rangle = \|P_{|T|}(A)x\|^2.$$

Hence Theorem 2.44 (ii) implies $\dim_{\mathcal{R}(G)} \operatorname{im} P_{|T|}(A) > 0$ which is equivalent to $0 < \operatorname{tr}_{\mathcal{R}(G)} P_{|T|}(A) = \mu_{|T|}(A)$. □

So the equivalence class of basic measures for the positive part of a morphism T of finitely generated Hilbert modules has the canonical representative $\mu_{|T|}$. The ℓ^2-version of a "determinant up to kernel" is now captured by the following definition. Let us set $\sigma(|T|)^+ = \sigma(|T|) \setminus \{0\}$.

Definition 5.42 The *Fuglede–Kadison determinant* of $T : H \to K$ is

$$\operatorname{det}_{\mathcal{R}(G)} T = \exp\left(\int_{\sigma(|T|)^+} \log \mathrm{d}\mu_{|T|}\right).$$

The above Lebesgue integral is always defined because the positive part of the logarithm function is bounded on $\sigma(|T|)^+ \subseteq (0, \|T\|]$. It might happen, though, that $|T|$ has so much spectral mass around zero that the integral has value $-\infty$. In this case, we can and will set $\operatorname{det}_{\mathcal{R}(G)} T = \exp(-\infty) = 0$. But we want to say that T is of *determinant class* if $\operatorname{det}_{\mathcal{R}(G)} T > 0$, or in other words, if log is $\mu_{|T|}$-integrable on $\sigma(|T|)^+$. Also be aware that the zero operator has all its spectral mass at the eigenvalue zero which is excluded from integration. Thus $\operatorname{det}_{\mathcal{R}(G)} 0 = \exp(0) = 1$.

To understand why this definition gives a notion of determinant, it is advisable to decode it in case the group G is trivial so that $H \cong \mathbb{C}^k$. In that case, $\sigma(|T|)^+$ consists of the finitely many positive eigenvalues of $|T|$, also known as the positive *singular values* of T, and Proposition 5.15 says that $\mu_{|T|}(\{\lambda\})$ is the multiplicity of $\lambda \in \sigma(|T|)^+$. It follows that $\det_{\mathcal{R}(G)} T$ is the product of the positive singular values of T repeating them according to multiplicities. Similarly, for a finite group G, the Fuglede–Kadison determinant $\det_{\mathcal{R}(G)} T$ is the $|G|$-th root of the product of positive singular values of T raised with multiplicities.

In any case, the inequality $\lambda_1^{m_1} \cdots \lambda_s^{m_s} \geq 1$ from (5.5) can now be restated as $\det_{\mathcal{R}(G/G_i)}(\cdot D_i) \geq 1$. We make the bold claim that this should not only be true for finite groups but in fact for all groups.

Conjecture 5.43 (Determinant Conjecture) *Let G be any group and let $A \in M(k, l; \mathbb{Z}G)$ be any matrix. Then*

$$\det_{\mathcal{R}(G)} \left((\ell^2 G)^k \xrightarrow{\ \cdot A\ } (\ell^2 G)^l \right) \geq 1.$$

Just like (5.5), the determinant conjecture, if true, yields a uniform logarithmic bound for the positive spectral distribution function of $\cdot A$ and thus provides the missing part for the approximation conjecture.

Proposition 5.44 (Logarithmic Bound II) *Let $T : H \to K$ be a morphism of finitely generated Hilbert $\mathcal{L}(G)$-modules. Suppose $\dim_{\mathcal{R}(G)} H \leq k$ and $\det_{\mathcal{R}(G)} T \geq 1$. Then for all $\lambda \in (0, 1)$, we have*

$$\mu_{|T|}((0, \lambda)) \leq \frac{k \log \|T\|}{|\log \lambda|}.$$

Proof The proof is the continuous version of the argument given in Proposition 5.18. Indeed, $\det_{\mathcal{R}(G)} T \geq 1$ gives

$$0 \leq \int_{\sigma(|T|)^+} \log \mathrm{d}\mu_{|T|} = \int_{0+}^{\lambda^-} \log \mathrm{d}\mu_{|T|} + \int_{\lambda}^{\|T\|} \log \mathrm{d}\mu_{|T|} \leq$$

$$\leq \log \lambda \cdot \mu_{|T|}((0, \lambda)) + \log \|T\| \cdot \mu_{|T|}([\lambda, \|T\|]) \leq$$

$$\leq \log \lambda \cdot \mu_{|T|}((0, \lambda)) + \log \|T\| \cdot k. \qquad \square$$

We are now in the position to state and prove that the determinant conjecture implies the approximation conjecture in the following sense.

Theorem 5.45 *Let G be a group and let $(G_i)_{i \in I}$ be a residual system. If each group G/G_i satisfies the determinant conjecture, then G and $(G_i)_{i \in I}$ satisfy the approximation conjecture 5.39.*

Proof Fix $A \in M(k, l; \mathbb{Q}G)$ and let c be the l.c.m. of the denominators of the coefficients in the entries of A. The matrix $D = c^2 A^* A \in M(k, k; \mathbb{Z}G)$ and the reductions $D_i \in M(k, k; \mathbb{Z}(G/G_i))$ are positive and have the same kernels as A and A_i. Thus we have to show $\mu._D(\{0\}) = \lim_{i \in I} \mu._{D_i}(\{0\})$. As discussed above, this follows once we prove that for all $i \in I$ and all $\lambda \in (0, 1)$, we have

$$\mu._{D_i}((0, \lambda)) \leq \frac{k \log d}{|\log \lambda|}.$$

But if the determinant conjecture is true for each G/G_i, then this is implied by Proposition 5.44 and the inequality $\| \cdot D_i \| \leq k^2 \cdot \|D\|_1 := d$. $\square$

This theorem draws the attention from the approximation conjecture toward the determinant conjecture. We shall now endeavor to prove the determinant conjecture for a reasonable class of groups. We start with an entirely measure theoretic consideration. Let X be a (not necessarily compact) metrizable space and suppose that a net $(\mu_i)_{i \in I}$ of finite Borel measures on X *weakly converges* to a finite Borel measure μ, meaning that

$$\lim_{i \in I} \int f \, d\mu_i = \int f \, d\mu$$

for all *bounded* continuous functions $f \in C_b(X, \mathbb{R})$. Then for possibly unbounded nonnegative functions, we still get the following inequality.

Lemma 5.46 *For every continuous function* $f : X \to [0, \infty)$, *we have*

$$\liminf_{i \in I} \int f \, d\mu_i \geq \int f \, d\mu.$$

As usual in these contexts, integrals are allowed to take the value ∞. With a sequence of measures instead of a net, the lemma is given as [43, Aufgabe 4.13, p. 409]. For the convenience of the reader we provide a proof.

Proof For all $i \in I$ and all $n \in \mathbb{N}$, we have

$$\int \min(f, n) \, d\mu_i \leq \int f \, d\mu_i$$

by monotonicity of the integral. Taking the limit inferior over $i \in I$ gives

$$\int \min(f, n) \, d\mu \leq \liminf_{i \in I} \int f \, d\mu_i$$

for all $n \in \mathbb{N}$ by weak convergence of the measures μ_i to μ. Taking the limit $n \to \infty$ completes the proof by the monotone convergence theorem. $\square$

The inequality of Lemma 5.46 implies the following inequality for the determinant of a matrix and of its reductions.

Proposition 5.47 *Let G be a group with residual system $(G_i)_{i \in I}$ such that each group G/G_i satisfies the determinant conjecture. Then for every $A \in M(k, l; \mathbb{Q}G)$, we have*

$$\det_{\mathcal{R}(G)} \cdot A \geq \limsup_{i \in I} \det_{\mathcal{R}(G/G_i)} \cdot A_i.$$

Proof Since for $c > 0$, we have $\det_{\mathcal{R}(G)}(\cdot cA) = c^k \det_{\mathcal{R}(G)}(\cdot A)$ and similarly for cA_i, we can multiply A with the l.c.m. of the denominators, if need be, to assume $A \in M(k, l; \mathbb{Z}G)$. The proof of Proposition 5.16 works equally well when the groups (G_i) form a residual system instead of a residual chain. So applying this proposition to $\cdot A^* A$ and $\cdot A_i^* A_i$, we see that the spectral measures $\mu_{|\cdot A_i|^2}$ converge weakly to $\mu_{|\cdot A|^2}$ on the closed interval $[0, a^2]$ with $a^2 = k^2 \cdot \|A^* A\|_1$. Since for all $f \in C([0, a], \mathbb{R})$, we have

$$\int f(x) \, d\mu_{|\cdot A|}(x) = \int f(\sqrt{x}) \, d\mu_{|\cdot A|^2}(x),$$

and similarly for A_i, the net of spectral measures $\mu_i = \mu_{|\cdot A_i|}$ converges weakly to $\mu = \mu_{|\cdot A|}$ on $[0, a]$. As we assume that G/G_i satisfies the determinant conjecture, Proposition 5.44 and the inequality $\| \cdot A_i \| \leq a$ give

$$\mu_i((0, \lambda)) \leq \frac{k \log a}{|\log \lambda|} \tag{5.10}$$

for all $\lambda \in (0, 1)$ and all $i \in I$. We now show that this implies that μ_i converges also weakly to μ on the open interval $(0, 1]$. Indeed, let $f \in C_b((0, 1], \mathbb{R})$ be a continuous function bounded by $|f| \leq C$. Then splitting the domain of integration as $(0, 1] = (0, \lambda) \cup [\lambda, 1]$, we obtain for each $i \in I$ the estimate

$$-C\mu_i((0, \lambda)) + \int_\lambda^1 f \, d\mu_i \leq \int_{0+}^1 f \, d\mu_i \leq C\mu_i((0, \lambda)) + \int_\lambda^1 f \, d\mu_i.$$

Together with (5.10), this gives

$$-\frac{C'}{|\log \lambda|} + \int_\lambda^1 f \, d\mu_i \leq \int_{0+}^1 f \, d\mu_i \leq \frac{C'}{|\log \lambda|} + \int_\lambda^1 f \, d\mu_i$$

with $C' = Ck \log a$. The function f can clearly be extended continuously from $[\lambda, 1]$ to $[0, a]$, so $\mu_i \to \mu$ weakly on $[\lambda, 1]$. Hence first taking $\liminf_{i \in I}$ and

$\limsup_{i \in I}$, respectively, then forming the limit $\lambda \to 0^+$ we obtain

$$\liminf_{i \in I} \int_{0^+}^1 f \, d\mu_i \geq \int_{0^+}^1 f \, d\mu \geq \limsup_{i \in I} \int_{0^+}^1 f \, d\mu_i,$$

which gives the asserted weak convergence of μ_i to μ on $(0, 1]$. Now we have

$$\limsup_{i \in I} \log \det_{\mathcal{R}(G/G_i)} \cdot A_i = -\liminf_{i \in I} \int_{0^+}^1 (-\log) \, d\mu_i + \limsup_{i \in I} \int_1^a \log \, d\mu_i.$$

As μ_i converges weakly to μ both on $(0, 1]$ and on $[1, a]$, Lemma 5.46 gives

$$\limsup_{i \in I} \log \det_{\mathcal{R}(G/G_i)} \cdot A_i \leq -\int_{0^+}^1 (-\log) \, d\mu + \int_1^a \log \, d\mu = \log \det_{\mathcal{R}(G)} \cdot A.$$

Finally, the logarithm function is monotone increasing, therefore commutes with $\limsup_{i \in I}$. This completes the proof.   □

Remark 5.48 The statement that in fact we should have

$$\det_{\mathcal{R}(G)} \cdot A = \lim_{i \in I} \det_{\mathcal{R}(G/G_i)} \cdot A_i$$

for any group G, any residual system (G_i), and any matrix $A \in M(k, l; \mathbb{Q}G)$ goes by the name *determinant approximation conjecture*, neither to be confused with the *determinant conjecture* nor with the *approximation conjecture*... At the time of writing, the determinant approximation conjecture is wide open. The inequality opposite to Proposition 5.47

$$\det_{\mathcal{R}(G)} \cdot A \leq \liminf_{i \in I} \det_{\mathcal{R}(G/G_i)} \cdot A_i \tag{5.11}$$

turns out to be surprisingly hard to establish. It seems that as of now, it is only known for virtually cyclic G, see [159] and [117, Lemma 13.53, p. 478]. Even in this case, a technical result from Diophantine approximation enters the proof, namely a precursor of Baker's famous theorem on linear forms in logarithms. In [87, Section 4] the reader can find a short excursion to this beautiful part of transcendental number theory in which the technical result is stated as Theorem 15. We will revisit inequality (5.11) at the end of Sect. 6.5 in Chap. 6.

   If matrices are allowed to have coefficients in $\mathbb{C}G$ instead of $\mathbb{Q}G$, the determinant approximation conjecture becomes *wrong* even in the case $G = \mathbb{Z}$ and $k = l = 1$. A counterexample is presented in [117, Example 13.69, p. 481]. For amenable G, Li–Thom [103, Theorem 1.4] show that Fuglede–Kadison determinants can be approximated by the determinants of the operators obtained by restricting and projecting to subspaces $\ell^2(F)^k$ for $F \subset G$ finite.

Theorem 5.49 *Let G be a group with residual system $(G_i)_{i \in I}$. If each quotient group G/G_i satisfies the determinant conjecture, then so does G.*

Proof Immediate from Proposition 5.47. □

Corollary 5.50 *Residually finite groups satisfy the determinant conjecture.*

Proof By (5.5), finite groups satisfy the determinant conjecture. □

Combining Theorem 5.45 with Corollary 5.50 gives the following result.

Theorem 5.51 *Let G be a group with residual system $(G_i)_{i \in I}$ such that each quotient group $(G/G_i)_{i \in I}$ is residually finite. Then G and $(G_i)_{i \in I}$ satisfy the approximation conjecture.*

This theorem finally improves Lück's approximation theorem from finite quotient groups to residually finite quotient groups. Be aware, however, that a residually residually finite group is residually finite. So Theorem 5.51 does not yet apply to a more general class of groups G than Lück's approximation theorem does. It does however apply to more general systems of subgroups $(G_i)_{i \in I}$ in a residually finite group G.

We now fulfill our promise from the end of Sect. 3.5 in Chap. 3 and illustrate how the approximation conjecture gives further insight into the Atiyah conjecture by Schick's strategy in Theorem 5.40. Knowing or accepting the Atiyah conjecture 3.30 for elementary amenable groups (which are "close" to being abelian), we can conclude it for free groups (which are far from being abelian).

Theorem 5.52 *The Atiyah conjecture for elementary amenable torsion-free groups implies the Atiyah conjecture for free groups in case $R = \mathbb{Q}$.*

Proof Similarly to the argument below Conjecture 5.39, a matrix over the rational group ring of a free group involves only finitely many words in finitely many letters from a free generating set. Hence we can replace the free group by the free subgroup generated by these finitely many letters. Thus it suffices to show the theorem for the free group on n letters $G = F_n$.

The lower central series $(G_i)_{i=0}^{\infty}$ of G, recursively defined by $G_0 = G$ and $G_{i+1} = [G, G_i]$, is a residual chain in G. The quotient groups G/G_i are torsion-free nilpotent [67, Chapter 11]. In particular, they are elementary amenable, hence satisfy the Atiyah conjecture 3.30 with $R = \mathbb{Q}$ by assumption. Finitely generated nilpotent groups are moreover residually finite as shown in [75]. Theorem 5.51 and Theorem 5.40 complete the proof. □

In the remainder of this chapter, we inform on further developments in this circle of ideas, only hinting at proofs as we feel inclined to do so. For the determinant conjecture, we can state a surprisingly encompassing result.

Theorem 5.53 (Elek–Szabó [42], Schick [157]) *The class of groups satisfying the determinant conjecture contains all sofic groups and is closed under limits and colimits of directed systems, subgroups, and amenable extensions.*

Note that limits of groups are typically uncountable but any matrix is supported in a finitely generated subgroup so that this is no issue. Schick [157] showed that the property of satisfying the determinant conjecture has the asserted closure properties. In this context, being closed under *amenable extensions* is meant a little more general than just saying that G satisfies the conjecture if G has a normal subgroup N such that G/N is amenable and N satisfies the conjecture; see [157, Definition 1.12 and 1.13] for the precise statement. Since the conjecture holds for the trivial group, Schick's result already shows that all residually amenable groups satisfy the conjecture. Elek and Szabó [42] proved subsequently that the determinant conjecture holds for the humongous class of *sofic* groups, a notion due to Gromov [65] and named so by Weiss [171] that simultaneously generalizes amenability and residual finiteness. At the time of writing, no example of a non-sofic group is known. However, experts seem to believe they exist and constructing a matrix $A \in M(k, l; \mathbb{Z}G)$ with $\det_{\mathcal{R}(G)} \cdot A < 1$ might be a strategy to find one.

What can be said about the approximation conjecture 5.39 if we allow more general coefficients? Asking this might not have an immediate topological gain, but it is still of algebraic interest as we can again draw conclusions on the Atiyah conjecture which in turn has consequences for Kaplansky's conjecture with more general coefficient fields.

Conjecture 5.54 (Approximation Conjecture with Coefficients in K) Let G be a group with residual system $(G_i)_{i \in I}$ and let $K \subset \mathbb{C}$ be any subfield. Then for all $A \in M(k, l; KG)$ with reductions $A_i \in M(k, l; K(G/G_i))$, we get

$$\dim_{\mathcal{R}G} \ker(\ell^2 G^k \xrightarrow{\cdot A} \ell^2 G^l) = \lim_{i \in I} \dim_{\mathcal{R}(G/G_i)} \ker(\ell^2 (G/G_i)^k \xrightarrow{\cdot A_i} \ell^2 (G/G_i)^l).$$

If $K = \overline{\mathbb{Q}}$ is the field of algebraic numbers, then a matrix $A \in M(k, l; \overline{\mathbb{Q}}G)$ has in fact entries in FG where F is a finite Galois extension of $\mathbb{Q}$. Multiplying with a rational integer, if need be, we may assume that $A \in M(k, l; O_F G)$ where O_F is the ring of integers of F. With similar "bootstrapping" methods as before, this integrality can be exploited to show that Fuglede–Kadison determinants are bounded from below by a positive constant if the groups G/G_i are obtained from the trivial group by successive application of the operations listed in Theorem 5.53. Similar to Proposition 5.44, we obtain a logarithmic spectral bound which suffices to conclude the conjecture. This method is due to Dodziuk et al. [37, Theorem 3.7]. Jaikin-Zapirain sketches in [80, Section 10.4] how to incorporate the methods of Elek–Szabó to conclude that G and (G_i) also satisfy the approximation conjecture with coefficients in $\overline{\mathbb{Q}}$ if each quotient G/G_i is sofic. But $\overline{\mathbb{Q}}$-coefficients take the method of finding lower bounds for determinants to the limit. Once transcendental numbers occur in the matrix, the Fuglede–Kadison determinants of the reduced matrices can converge to zero [80, Section 10.3].

Hence the most general case $K = \mathbb{C}$ calls for new techniques. If G is amenable, then so are all the quotients G/G_i and the approximation conjecture with coefficients in $\mathbb{C}$ was proven by Elek [41], see also [141]. The breakthrough is however due to Jaikin-Zapirain who recently pioneered an innovative algebraic approach [81] to address this question.

Theorem 5.55 (Jaikin-Zapirain [79]) *Let G be a group with a residual system $(G_i)_{i \in I}$ such that each quotient group G/G_i is sofic. Then G and $(G_i)_{i \in I}$ satisfy the approximation conjecture with coefficients in $\mathbb{C}$.*

Moreover, Jaikin-Zapirain proves the approximation conjecture with coefficients in $\mathbb{C}$ for free groups with arbitrary residual systems [79]. It was long known that the Kaplansky conjecture with $R = \overline{\mathbb{Q}}$ and $R = \mathbb{C}$ are actually equivalent, see for instance [37, Proposition 5.1]. For sofic groups, Theorem 5.55 has the same striking consequence on the Atiyah conjecture.

Theorem 5.56 (Jaikin-Zapirain [79]) *If G is sofic, then the Atiyah conjecture with $R = \overline{\mathbb{Q}}$ is equivalent to the Atiyah conjecture with $R = \mathbb{C}$.*

This is a particularly convenient theorem because as opposed to Linnell's theorem 3.33, the more recent results on the Atiyah conjecture were obtained for $R = \overline{\mathbb{Q}}$ rather than $\mathbb{C}$. For example, the authors of [37] applied the approximation methods sketched above similarly as in Theorem 5.40 to prove the Atiyah conjecture with $R = \overline{\mathbb{Q}}$ for the following class of groups.

Definition 5.57 Let $\mathcal{D}$ be the smallest nonempty class of groups that

- contains every torsion-free group G for which there exists an epimorphism $p \colon G \to A$ onto an elementary amenable group A such that $p^{-1}(H) \in \mathcal{D}$ for every finite subgroup $H \leq A$.
- is closed under taking limits, colimits, and subgroups.

So in particular, residually torsion-free solvable groups lie in $\mathcal{D}$. Incorporating additional work of Farkas and Linnell [45], Linnell and Schick [106], Schreve [161], and Jaikin-Zapirain and López-Álvarez [82], Theorem 5.56 implies the following extensive result on the Atiyah conjecture [81, Corollary 1.2].

Theorem 5.58 *The Atiyah conjecture with $R = \mathbb{C}$ holds for groups in $\mathcal{D}$, Artin braid groups, finite extensions of the fundamental group of a compact special cube complex, torsion-free p-adic analytic pro-p groups, and locally indicable groups.*

We will not define or explain the additional classes of groups occurring in this theorem. But let us mention that the result on analytic pro-p groups implies that every finitely generated *linear* group over a field of characteristic zero has a finite index subgroup satisfying the conjecture. Unfortunately, it is not known in general if the Atiyah conjecture passes to finite index overgroups. Partial results on this question were however used in the proof for braid groups and virtually cocompact special groups.

As another side remark Wise [173, Theorem 1.4], showed that one relator groups with torsion are virtually cocompact special, hence satisfy the Atiyah conjecture. If the letters in the relator word of a one relator group occur with only nonnegative powers and the abelianization of the group is torsion-free, then the group itself is torsion-free and residually in Linnell's class C [156, Example 4.1]. Therefore it lies in $\mathcal{D}$ and likewise satisfies the Atiyah conjecture. Finally, Jaikin-Zapirain and López-Álvarez showed most recently that the Atiyah conjecture holds for locally indicable groups. This includes the torsion-free one relator groups by work of Brodskii [27]. For more information on all these recent developments in the approximation theory of ℓ^2-Betti numbers, we recommend the survey articles [80, 96].

Chapter 6
Torsion Invariants

Let us step back and take a look on what we have achieved so far. The starting point was to consider the n-th Betti number $b_n(X)$ of a finite CW complex X, which is defined as $\mathrm{rank}_{\mathbb{Z}} H_n(X)_{\text{free}}$ where $H_n(X) \cong H_n(X)_{\text{free}} \oplus H_n(X)_{\text{tors}}$ is the decomposition of the n-th integral homology into free and torsion part. We introduced the n-th ℓ^2-Betti number $b_n^{(2)}(\widetilde{X})$ as the ℓ^2-counterpart to $b_n(X)$ and Lück's approximation theorem says that it can be recovered asymptotically from the Betti numbers $b_n(\overline{X}_i)$ of finite coverings $\overline{X}_i \to X$ if $\pi_1 X$ is residually finite. While this is a very satisfying theory, it came at the cost of completely discarding torsion in homology.

Torsion in homology is however an object of utmost interest so it makes sense to ask for a theory along the above lines that would define an ℓ^2-invariant of $\widetilde{X}$ that could asymptotically be recovered from the finite groups $H_n(\overline{X}_i)_{\text{tors}}$. The good news is that such an invariant exists. It is called ℓ^2-*torsion*, and we have a clean conjecture stating how and under what conditions it can be recovered from the groups $H_n(\overline{X}_i)_{\text{tors}}$. The bad news is that the conjecture is entirely open. Nevertheless, it is instructive and worthwhile to expose the difficulties of the conjecture so that at the end of this chapter the reader has an impression of the state of the art in this circle of question which has attracted massive research efforts from various fields, including 3-manifold theory and cohomology of arithmetic groups.

6.1 Reidemeister Torsion

To begin with, we will present the classical invariant from which ℓ^2-torsion arises in the ℓ^2-setting. It goes by the name of *Reidemeister torsion* or *Reidemeister–Franz torsion*. To motivate the definition, consider the 3-sphere S^3 as the unit sphere in $\mathbb{C}^2$ and let p and q be coprime integers. We define a free action of the cyclic group $\mathbb{Z}/p$ on S^3 by saying that the generator of $\mathbb{Z}/p$ moves the point $(z_1, z_2) \in S^3 \subseteq \mathbb{C}$ to

© Springer Nature Switzerland AG 2019
H. Kammeyer, *Introduction to ℓ^2-invariants*, Lecture Notes in Mathematics 2247,
https://doi.org/10.1007/978-3-030-28297-4_6

the point $(e^{2\pi iq/p}z_1, e^{2\pi i/p}z_2)$. The quotient space $L(p,q) = S^3 / \mathbb{Z}/p$ is hence a three-dimensional manifold.

Definition 6.1 The manifold $L(p,q)$ is called a *lens space* of type (p,q).

Let us compute the homotopy and homology groups of $L(p,q)$. Since S^3 is simply-connected and $\mathbb{Z}/p$ acts freely, we have $\pi_1(L(p,q)) \cong \mathbb{Z}/p$. The higher homotopy groups $\pi_k(L(p,q))$ for $k \geq 2$ are equal to the homotopy groups $\pi_k S^3$ because S^3 is the universal covering of $L(p,q)$. With the help of Poincaré duality and the universal coefficient theorem one readily verifies that the homology of $L(p,q)$ is given by $H_0(L(p,q)) \cong \mathbb{Z}$, $H_1(L(p,q)) \cong \mathbb{Z}/p$, $H_2(L(p,q)) \cong 0$ and $H_3(L(p,q)) \cong \mathbb{Z}$. The upshot of this is that the elementary algebraic topology of $L(p,q)$ does not see the integer q. Nevertheless, the lens space $L(5,1)$ is not homotopy equivalent to the lens space $L(5,2)$ and—even worse—the lens space $L(7,1)$ is homotopy equivalent to $L(7,2)$ but they are not homeomorphic! But how does one even prove that? All common invariants in topology (including refinements like cup products in cohomology) are homotopy invariants and thus will not be able to distinguish $L(7,1)$ from $L(7,2)$. An object which is however not a homotopy invariant of a CW complex X is the cellular chain complex $C_*(X; \mathbb{R})$, for example with coefficients in $\mathbb{R}$. Or more generally, if X is a G-CW complex, one could consider the cellular chain complex $C_*(X; V) = V \otimes_{\mathbb{Z}G} C_*(X)$ for any finite-dimensional representation V of G over $\mathbb{R}$. Of course, $C_*(X; V)$ is not even invariant under refinements of the cell structure so that it is hardly a useful thing to work with directly. But instead of taking homology, there is another way to extract useful information hidden in $C_*(X; V)$ even if—or better *especially if*—$C_*(X; V)$ has trivial homology. To do so, let us first advertise an intuitive picture to think about chain complexes.

You grab a stack of beer coasters, allowed to be of varying sizes, and place half of the coasters side by side on the table without overlaps so that some gap remains in between any two adjacent coasters. Afterwards, you use the other half of the coasters to cover the gaps so that any gap between two adjacent upper coasters lies above some particular lower coaster.

What's that got to do with chain complexes? The lower beer coasters represent the even chain modules C_{2*}, the upper ones correspond to the odd chain modules C_{2*+1}. The overlaps between upper and lower coasters determine how much of each chain module is transported to the next chain module by the differential.

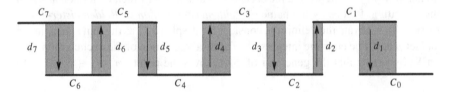

Requiring that neither the upper nor the lower coasters overlap among themselves thus translates precisely to the chain complex condition $\operatorname{im} d_{*+1} \subseteq \ker d_*$. Accordingly, the upper gaps represent the even homology groups and the lower gaps account for the odd homology groups of the chain complex. The two extreme cases would be the picture

where all differentials are zero and thus the gaps (homology) are as big as the coasters (chain modules) and the picture

where the chain complex is *exact* (or *acyclic*), $\operatorname{im} d_{*+1} = \ker d_*$, and thus there are no gaps (no homology). In the latter case, visually $C_{\text{odd}} = \bigoplus C_{2*+1}$ is isomorphic to $C_{\text{even}} = \bigoplus C_{2*}$.

To see this isomorphism formally, we assume that C_* consists of (finitely many) finite dimensional $\mathbb{R}$-vector spaces as in the example $C_* = C_*(X; V)$. Then each C_* is automatically free and the condition $H_*(C_*) = 0$ ensures that (C_*, d_*) is *contractible*: there exists a *chain contraction* $\gamma_* : C_* \to C_{*+1}$ satisfying $\gamma_{*-1}d_* + d_{*+1}\gamma_* = \operatorname{id}_{C_*}$.

Proposition 6.2 *The map* $d_{2*+1} + \gamma_{2*+1} \colon C_{\text{odd}} \to C_{\text{even}}$ *is an isomorphism of vector spaces.*

Proof The composition $(d_{2*+1} + \gamma_{2*+1})(d_{2*} + \gamma_{2*})$ and the reverse composition $(d_{2*}+\gamma_{2*})(d_{2*+1}+\gamma_{2*+1})$ are unipotent endomorphisms and in particular invertible. $\square$

So the map $d_{2*+1}+\gamma_{2*+1} \colon C_{\text{odd}} \to C_{\text{even}}$ is represented by a nonsingular square matrix as soon as we fix a basis for all the vector spaces C_*.

Proposition 6.3 *The number* $\det(d_{2*+1}+\gamma_{2*+1}) \in \mathbb{R}^*$ *is independent of the choice of the chain contraction* γ_*.

Proof Let $\delta \colon C_* \to C_{*+1}$ be another chain contraction. Set $\mu_* = (\gamma_{*+1} - \delta_{*+1})\delta_*$. Then both $(\operatorname{id} + \mu_{2*+1})$ and the composition

$$(d_{2*+1} + \gamma_{2*+1})(\operatorname{id} + \mu_{2*+1})(d_{2*} + \delta_{2*})$$

are unipotent, thus the number $\det(d_{2*+1}+\gamma_{2*+1}) = \det(d_{2*}+\delta_{2*})^{-1}$ is independent of γ (and δ). $\square$

Definition 6.4 The *Reidemeister torsion* of C_* is given by

$$\rho(C_*) = |\det(d_{2*+1} + \gamma_{2*+1})| \in \mathbb{R}^{>0}.$$

Other authors leave out the absolute value [149] or square the determinant instead [31]. While Reidemeister torsion is independent of the chain contraction, it depends decisively on the chosen bases. In fact, if we replace the basis for each C_p by a new one, and if A_p denotes the change of basis matrix, then the new Reidemeister torsion differs from the old by the factor

$$\cdots |\det A_2|^{-1} |\det A_1| \, |\det A_0|^{-1} |\det A_{-1}| \cdots .$$

This means, however, that Reidemeister torsion remains unchanged if all change of basis matrices are orthogonal. To obtain a well-defined invariant, it is thus enough to specify the bases up to orthogonal transformations, or in other words to fix an *inner product* on each C_n. This makes computing Reidemeister torsion particularly easy because the inner product gives a convenient, canonical chain contraction. To see that, consider the orthogonal decomposition

$$C_n = (\ker d_n) \oplus (\ker d_n)^{\perp} = \operatorname{im} d_{n+1} \oplus \operatorname{im} d_n^*.$$

The differential d_n restricts to an isomorphism

$$d_n^{\perp} = d_{n \,|\, \operatorname{im} d_n^*} : \operatorname{im} d_n^* \to \operatorname{im} d_n$$

so that a chain contraction with respect to the above decomposition is given by $\gamma_n = \begin{pmatrix} 0 & 0 \\ d_{n+1}^{\perp}{}^{-1} & 0 \end{pmatrix}$. The isomorphism $d_{2*+1} + \gamma_{2*+1} : C_{\text{odd}} \to C_{\text{even}}$ is then given in block form as

$$d_{2*+1} + \gamma_{2*+1} = \begin{pmatrix} \ddots & & & & \\ & 0 & d_1^{\perp} & & \\ & d_2^{\perp}{}^{-1} & 0 & & \\ & & & 0 & d_{-1}^{\perp} \\ & & & d_0^{\perp}{}^{-1} & 0 \\ & & & & & \ddots \end{pmatrix}$$

The inner products on the various C_n add up to inner products on C_{odd} and C_{even}. Therefore, we obtain a positive endomorphism $|d_{2*+1} + \gamma_{2*+1}|$ acting on C_{odd} which is defined by requiring that it have the same eigenspace decomposition as $(d_{2*+1} + \gamma_{2*+1})^*(d_{2*+1} + \gamma_{2*+1})$ but with square rooted eigenvalues. In block form it is given by

$$|d_{2*+1} + \gamma_{2*+1}| = \begin{pmatrix} \ddots & & & & \\ & |d_2|^{\perp}{}^{-1} & & & \\ & & |d_1|^{\perp} & & \\ & & & |d_0|^{\perp}{}^{-1} & \\ & & & & |d_{-1}|^{\perp} \\ & & & & & \ddots \end{pmatrix}$$

and is nicely illustrated by the beer coaster picture without gaps. Here we used $|d_n^{\perp -1}| = |d_n^{\perp}|^{-1}$, because $|x^{-1}| = |x|^{-1}$ for all $x \in \mathbb{R}^*$, and $|d_n^{\perp}| = |d_n|^{\perp}$, because d_n and $|d_n|$ have the same kernel. It is clear that for any choice of orthonormal bases of C_n we have $|\det(d_{2*+1} + \gamma_{2*+1})| = \det|d_{2*+1} + \gamma_{2*+1}|$. Thus we have proven the following result.

Proposition 6.5 *Let (C_*, d_*) be a finite, acyclic chain complex of finite-dimensional real inner product spaces. Then the Reidemeister torsion (with respect to any collection of orthonormal bases of the C_n) is given by*

$$\rho(C_*) = \prod_{n \in \mathbb{Z}} \det |d_n|^{\perp(-1)^{n+1}}.$$

Let us now return to topology and consider a finite, free G-CW complex X, for instance the universal covering of $L(p, q)$. How do we obtain a chain complex from X that fits in our picture? Well, we pick an orthogonal representation $\varphi \colon G \to O(V)$ on some finite-dimensional real inner product space V with the property that the twisted chain complex $C_*(X; V) = V \otimes_{\mathbb{Z}G} C_*(X)$ is acyclic. Here V is turned into a $\mathbb{Z}G$-right module by setting $v \cdot g = \varphi(g^{-1})(v)$. Existence of such a representation must be checked case by case. Note however that the trivial representation $V = \mathbb{R}$ will never work because $C_*(X; \mathbb{R})$ will be infinite-dimensional unless G is finite and—what is worse—it is never acyclic because $H_0(X; \mathbb{R}) \cong \mathbb{R}\pi_0(X)$. Working with an orthogonal representation has the effect that we obtain an inner product on $C_*(X; V)$, defined as usual: Choosing a cellular basis for X gives an identification of $C_n(X)$ with the free $\mathbb{Z}G$-left module $(\mathbb{Z}G)^{k_n}$. This in turn gives an isomorphism of $\mathbb{R}$-vector spaces $C_n(X; V) \cong V^{k_n}$ which defines an inner product on $C_n(X; V)$. Had we chosen a different cellular basis, then the change of basis matrix of V^{k_n} would be a generalized permutation matrix with entries $\pm\varphi(g)$ and thus an orthogonal transformation of V^{k_n}. It follows that the inner product is independent of the choice of cellular basis.

Definition 6.6 The *Reidemeister torsion* of X with coefficients in V is given by $\rho(X; V) = \rho(C_*(X; V))$.

The discussion so far justifies that we did not mention any bases in the definition any more. Reidemeister and Franz employed their torsion invariant to give the complete homeomorphism classification of three-dimensional lens spaces. To be historically correct, they gave the PL homeomorphism classification which was later shown to be the same as the homeomorphism classification by Brody [28]. The result is that $L(p, q_1)$ is homeomorphic to $L(p, q_2)$ if and only if $q_1 = \pm q_2^{\pm 1}$ mod p. By means of the *torsion linking form*, one can see that $L(p, q_1)$ is homotopy equivalent to $L(p, q_2)$ if and only if either $q_1 q_2$ or $-q_1 q_2$ is a quadratic residue mod p.

Exercises

6.1.1 Consider three dimensional projective space $\mathbb{RP}^3$ (which can be interpreted as a certain lens space). Let V be the nontrivial one-dimensional orthogonal representation of $\mathbb{Z}/2\mathbb{Z}$. Show that $V \otimes_{\mathbb{Z}[\mathbb{Z}/2\mathbb{Z}]} C_*(\widetilde{\mathbb{RP}^3})$ is acyclic and compute the Reidemeister torsion $\rho(\mathbb{RP}^3; V)$ of $\mathbb{RP}^3$ with coefficients in V.

6.2 ℓ^2-Torsion of CW Complexes

Reidemeister torsion as just defined is not only an invariant of the finite, free G-CW complex X but in fact of a pair (X, V) consisting of X and an orthogonal G-representation V. A topologist might find this unfortunate because she is interested in properties of the space X, ideally without any outside influence. The only canonical choice of a finite-dimensional representation V for the possibly infinite group G would be the trivial representation $\mathbb{R}$—which never gives rise to an acyclic complex $C_*(X; V)$. However, once one exits the familiar ground of linear algebra to enter the realm of Hilbert modules, the situation is better. There is a canonical unitary representation of G: the right regular representation on $\ell^2 G$. Moreover, the resulting chain complex $C_*(X, \ell^2 G)$ is just the ℓ^2-chain complex $C_*^{(2)}(X)$ which is often ℓ^2-acyclic as we saw in various examples, including hyperbolic 3-manifolds and mapping tori. These observations pave the way for the definition of ℓ^2-*torsion*, the ℓ^2-version of Reidemeister torsion.

To translate the formula

$$\rho(C_*) = \prod_{n \in \mathbb{Z}} \det |d_n|^{\perp (-1)^{n+1}}$$

from Proposition 6.5 to the ℓ^2-setting, we spell out that the factors "$\det |d_n|^\perp$" are determinants of the positive part of a morphism of Euclidean spaces restricted to the orthogonal complement of the kernel. Hence the Fuglede–Kadison determinants $\det_{\mathcal{R}(G)} d_n^{(2)}$ provide the perfect ℓ^2-counterpart and we can right away give the following definition.

Definition 6.7 Let $(C_*^{(2)}, d_*^{(2)})$ be a chain complex of finitely many finitely generated Hilbert $\mathcal{L}(G)$-modules. Assume $C_*^{(2)}$ is of *determinant class*: each $d_n^{(2)}$ is of determinant class. Then the ℓ^2-*torsion* of $C_*^{(2)}$ is

$$\rho^{(2)}(C_*^{(2)}) = \sum_{n \in \mathbb{Z}} (-1)^{n+1} \log \det_{\mathcal{R}(G)} d_n^{(2)}.$$

Comparing to Proposition 6.5, you will have noticed that we have taken the logarithm so that ℓ^2-torsion can take any real value and not only positive values as Reidemeister torsion does. There is no mathematical necessity to do so but it has the welcome effect that in a moment we will get additive formulas instead of multiplicative ones and it also yields a more visible resemblance of ℓ^2-torsion and Euler characteristic as yet to be discussed. We discussed that as opposed to Reidemeister torsion, the transition from chain complexes to topology needs no additional input.

Definition 6.8 Let X be a finite, proper G-CW complex which is ℓ^2-acyclic and of determinant class. Then the ℓ^2-*torsion* of X is given by

$$\rho^{(2)}(X) = \rho^{(2)}(C_*^{(2)}(X)).$$

Here it is of course understood that X is of *determinant class* if $C_*^{(2)}(X)$ is. If X is free, this is automatic from the determinant conjecture 5.43 which we have proven for residually finite groups in Corollary 5.50. Since 3-manifold groups and lattices in semisimple Lie groups with finite center are residually finite, being of determinant class is granted in typical geometric situations. In fact, Theorem 5.53 says that being of determinant class is almost never an issue. Again, we will write $\rho^{(2)}(G \curvearrowright X)$ whenever it seems appropriate to emphasize the dependence on the group action. As usual after introducing a new notion, we list some properties.

Theorem 6.9 (Computation of ℓ^2-Torsion) *Assume that all occurring G-CW complexes are of determinant class.*

 (i) Homotopy invariance. *Suppose the finite, free, ℓ^2-acyclic G-CW complexes X and Y are G-homotopy equivalent and assume the determinant conjecture holds true for G. Then $\rho^{(2)}(X) = \rho^{(2)}(Y)$.*

 (ii) Additivity. *Let X be a G-CW pushout of finite, free G-CW complexes*

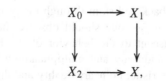

 where the upper map is an inclusion as G-invariant subcomplex. If three of the spaces are ℓ^2-acyclic, then so is the fourth and we have

$$\rho^{(2)}(X) = \rho^{(2)}(X_1) + \rho^{(2)}(X_2) - \rho^{(2)}(X_0).$$

 (iii) Multiplicativity. *Let $X \to Y$ be a d-sheeted covering of finite CW complexes such that $\widetilde{X}$ or $\widetilde{Y}$ is ℓ^2-acyclic. Then so is the other and*

$$\rho^{(2)}(\widetilde{X}) = d \cdot \rho^{(2)}(\widetilde{Y}).$$

(iv) Products. *Let X and Y be finite, free G- and H-CW complexes such that X is ℓ^2-acyclic. Then so is the $(G \times H)$-CW complex $X \times Y$ and*

$$\rho^{(2)}(X \times Y) = \rho^{(2)}(X) \chi(H \backslash Y).$$

(v) Poincaré duality. *Let X be a finite, free, ℓ^2-acyclic G-CW complex such that $G \backslash X$ is an orientable, closed, $2n$-manifold. Then $\rho^{(2)}(X) = 0$.*

(vi) Hyperbolic manifolds. *Suppose that $G \backslash X$ is a $(2n + 1)$-dimensional manifold. Assume either that it is closed and hyperbolic or has boundary and the interior carries a finite-volume hyperbolic metric. Then*

$$\rho^{(2)}(X) = (-1)^n C_n \, \mathrm{vol}(G \backslash X)$$

for a positive constant C_n that depends only on dimension.

The proof lies beyond the scope of this text because the somewhat intricate definition of ℓ^2-torsion effects that verifying these properties requires a considerably larger technical apparatus than was necessary for proving basic properties of ℓ^2-Betti numbers. Statement (vi) in particular has an involved proof which spreads over the papers [73, 110, 122].

It follows again from multiplicativity that $\rho^{(2)}(\widetilde{S^1}) = 0$. Similarly as in Chap. 3, Sect. 3.6.3, one can conclude from this that any connected, ℓ^2-acyclic, finite, free G-CW complex of determinant class with non-trivial S^1-action has vanishing ℓ^2-torsion. Thus Theorem 6.9 (vi) gives the second half of Theorem 1.3 from the introduction.

Corollary 6.10 *An odd dimensional closed hyperbolic manifold M does not permit any nontrivial action by the circle group.*

It is worthwhile to step back and skim through the properties of Theorem 6.9 with squinted eyes. In doing so, one should observe that the behavior of ℓ^2-torsion is strikingly reminiscent to the behavior of the Euler characteristic! In fact, homotopy invariance, additivity and multiplicativity hold true verbatim for ℓ^2-torsion and Euler characteristic. Poincaré duality and the values for hyperbolic manifolds, however, occur with shifted parity: Euler characteristic is zero for odd-dimensional manifolds and non-zero for even-dimensional hyperbolic manifolds. ℓ^2-torsion is zero for even-dimensional manifolds and non-zero for odd-dimensional hyperbolic manifolds. This brings us back to the beginning of this section where we said ℓ^2-torsion is a canonical invariant of spaces and thus should have a canonical interpretation: it is the *odd-dimensional cousin* of the Euler characteristic.

At this point, this might sound somewhat shaky but in Sect. 6.5 we discuss another deep manifestation of this principle in the context of *homology growth*. Beforehand, we include sections on ℓ^2-torsion of groups and ℓ^2-Alexander torsion of 3-manifolds in order to see some more examples and get acquainted with our new invariant.

6.3 ℓ^2-Torsion of Groups

ℓ^2-torsion is only defined for finite G-CW complexes. Because of Exercise 4.4.3, directly setting $\rho^{(2)}(G) = \rho^{(2)}(EG)$ would exclude any group G with torsion elements from the definition of ℓ^2-torsion. Let us therefore assume less restrictively that G virtually possesses a finite, ℓ^2-acyclic classifying space of determinant class. So we assume there is $H \leq G$ with $[G : H] < \infty$ and EH finite, ℓ^2-acyclic and of determinant class.

Definition 6.11 The ℓ^2-*torsion* of G is given by $\rho^{(2)}(G) = \frac{\rho^{(2)}(EH)}{[G:H]}$.

This is well-defined because if $H_1, H_2 \leq G$ are as above, then $H_1 \cap H_2$ is yet another allowed choice and multiplicativity (Theorem 6.9 (iii)) yields

$$\frac{\rho^{(2)}(EH_1)}{[G : H_1]} = \frac{\rho^{(2)}(E(H_1 \cap H_2))}{[G : H_1][H_1 : H_1 \cap H_2]} = \frac{\rho^{(2)}(E(H_1 \cap H_2))}{[G : H_1 \cap H_2]} = \frac{\rho^{(2)}(EH_2)}{[G : H_2]}.$$

Example 6.12 By Theorem 6.9 (vi), the fundamental group $G = \pi_1 M$ of an odd-dimensional hyperbolic manifold M has nonzero ℓ^2-torsion proportional to the volume.

Example 6.13 In dimension three, we have the following generalization. Suppose $G = \pi_1 M$ is an infinite fundamental group of a connected, compact, orientable, *irreducible* 3-manifold, meaning every embedded 2-sphere bounds a disk. Assume moreover that the boundary is either empty or a collection of tori. Then *Thurston geometrization* says one can cut M along embedded, *incompressible* tori into pieces each of which carries one out of eight *geometries* [7, Theorem 1.7.6]. A minimal choice of such tori is moreover unique up to isotopy. Here, a torus in M is *incompressible* if any embedded circle in the torus which is bounded by an embedded disk in M is already bounded by an embedded disk inside the torus. In this case we have

$$\rho^{(2)}(G) = -\frac{1}{6\pi} \sum_i \mathrm{vol}(M_i)$$

where the sum runs over the hyperbolic pieces [122, Theorem 0.7] as was conjectured in [113, Conjecture 7.7]. So $\rho^{(2)}(G) = 0$ if and only if M has no hyperbolic pieces in which case M is called a *graph manifold*.

Example 6.14 Example 6.12 generalizes in another way as follows. A Lie group G is called *semisimple* if the complexification of the Lie algebra of G has no nontrivial abelian ideal. Let G be a noncompact, semisimple linear Lie group and $\Gamma \leq G$ a *uniform lattice*: a discrete subgroup such that the quotient space $\Gamma \backslash G$ is compact. By Selberg's lemma [6], Γ possesses a finite index subgroup Λ which is torsion-free. Thus Λ intersects any fixed maximal compact subgroup $K \leq G$ trivially and therefore Λ acts freely on the *symmetric space* $X = G/K$. The symmetric space X

is moreover contractible and the *locally symmetric space* $\Gamma\backslash X$ is a closed manifold. Thus X possesses the structure of a contractible, free, finite Λ-CW complex and whence is a model for $E\Lambda$. If $\mathfrak{g}$ and $\mathfrak{k}$ are the Lie algebras of G and K, then the *deficiency* of G is the difference

$$\delta(G) = \operatorname{rank}_{\mathbb{C}} \mathfrak{g} \otimes \mathbb{C} - \operatorname{rank}_{\mathbb{C}} \mathfrak{k} \otimes \mathbb{C}.$$

It is a result of Borel [21] that Γ (equivalently Λ) is ℓ^2-acyclic if and only if $\delta(G) > 0$. In that case $\rho^{(2)}(\Gamma) = C(G, \mu) \cdot \mu(\Gamma\backslash G)$ where μ denotes both the Haar measure on G, see Sect. 4.5.5, and the induced G-invariant measure on $\Gamma\backslash G$. The constant $C(G, \mu)$ depends on G and μ only, and the product $C(G, \mu) \cdot \mu(\Gamma\backslash G)$ is of course independent of μ. By a result of Olbrich [139], we have $C(G, \mu) \neq 0$ if and only if $\delta(G) = 1$. For example $\delta(\mathrm{SO}^0(2n + 1, 1)) = 1$ in which case $\Lambda\backslash X$ is an odd-dimensional hyperbolic manifold as in Example 6.12. More generally, $\delta(\mathrm{SO}^0(p, q)) = 1$ if and only if $p \cdot q$ is odd. Up to isogeny, there is only one more simple Lie group of deficiency one: $G = \mathrm{SL}(3, \mathbb{R})$. Note that also $\delta(\mathrm{SL}(4, \mathbb{R})) = 1$ but this group is already accounted for because $\mathrm{SL}(4, \mathbb{R})$ is a finite covering space of $\mathrm{SO}^0(3, 3)$.

Example 6.15 Let G and K be as in the last example. Things become somewhat more involved if $\Gamma \leq G$ is a *non-uniform lattice*: a discrete subgroup such that the quotient space $\Gamma\backslash G$ is not compact but still has finite volume $\mu(\Gamma\backslash G)$. In that case X is nonetheless an $E\Lambda$ for any finite index, torsion-free subgroup $\Lambda \leq \Gamma$ but the Λ-CW structure is not finite. One can however construct a finite model of $E\Lambda$ from the manifold X by adding certain components at infinity to X. This construction goes by the name *Borel–Serre compactification* [22] and applies if Λ is an *arithmetic lattice*, meaning it is essentially given by the $\mathbb{Z}$-points of an algebraic group, see Sect. 6.6 for the precise definition. By a deep result of Margulis, to be presented on p. 152, assuming arithmeticity means no essential loss of generality provided G has "higher rank". The finite model of $E\Lambda$ is ℓ^2-acyclic if and only if $\delta(G) > 0$ just like in the uniform case. This follows from Gaboriau's proportionality principle in Sect. 4.5.5. As a consequence of a conjecture due to Lück et al. [123, Conjecture 1.2], for ℓ^2-torsion we should also have the same situation as in the uniform case: $\rho^{(2)}(\Gamma) \neq 0$ if and only if $\delta(G) = 1$. At the time of writing, this remains open in general. However, by inspecting closely the Borel–Serre compactification, one can conclude that $\rho^{(2)}(\Gamma) = 0$ if $\delta(G)$ is positive and even [84, Theorem 1.2].

The main example of a lattice in a semisimple Lie group is $\mathrm{SL}(k, \mathbb{Z})$. For this group the discussion boils down as follows. We have $b_1^{(2)}(\mathrm{SL}(2, \mathbb{Z})) = \frac{1}{12}$ because $\mathrm{SL}(2, \mathbb{Z}) \cong \mathbb{Z}/6 *_{\mathbb{Z}/2} \mathbb{Z}/4$. For $k \geq 3$, the group $\mathrm{SL}(k, \mathbb{Z})$ is ℓ^2-acyclic. Conjecturally, the values $\rho^{(2)}(\mathrm{SL}(3, \mathbb{Z}))$ and $\rho^{(2)}(\mathrm{SL}(4, \mathbb{Z}))$ are non-zero whereas $\rho^{(2)}(\mathrm{SL}(k; \mathbb{Z})) = 0$ for $k \geq 5$. But this is only known if $k = 1$ or $2 \bmod 4$. So at least we know that $\rho^{(2)}(\mathrm{SL}(5, \mathbb{Z})) = \rho^{(2)}(\mathrm{SL}(6, \mathbb{Z})) = 0$.

In addition to ℓ^2-torsion of groups one can also define ℓ^2-*torsion of automorphisms* of groups. To this end, we recall from Exercise 4.5.1 that if G has a finite model for BG, then for every automorphism $\varphi \in \mathrm{Aut}(G)$, the group $G \rtimes_\varphi \mathbb{Z}$ has a finite model which is ℓ^2-acyclic.

Definition 6.16 The ℓ^2-torsion of the automorphism $\varphi \in \mathrm{Aut}(G)$ is given by $\rho^{(2)}(\varphi) = \rho^{(2)}(G \rtimes_\varphi \mathbb{Z})$.

This invariant has many interesting properties and values but only recently has it gained attention in the literature, in particular so if G is free [30]. One can easily see that two automorphisms have equal ℓ^2-torsion if they differ by an inner automorphism so that each element $\gamma \in \mathrm{Out}(F_n)$ has well-defined ℓ^2-torsion. Some of these elements can be represented by self-homeomorphisms of a punctured surface so that the ℓ^2-torsion gives the hyperbolic volume of the corresponding mapping torus as in Example 6.13. In any case, it would be interesting to characterize the countable subset $\rho^{(2)}(\mathrm{Out}(F_n)) \subset \mathbb{R}$ of the real numbers. Here the notation "$\rho^{(2)}(\mathrm{Out}(F_n))$" should not be confused with the ℓ^2-torsion of the group $\mathrm{Out}(F_n)$: the latter is not defined because for the (rational) Euler characteristic, we have $\chi(\mathrm{Out}(F_n)) < 0$ for all $n \geq 2$ as was most recently announced by Borinsky and Vogtmann. In fact, D. Gaboriau announced that $b^{(2)}_{2n-3}(\mathrm{Out}(F_n)) > 0$ for $n \geq 2$ and this is the highest possible degree with non-vanishing ℓ^2-homology because the barycentric subdivision of the *spine of outer space* [168] is a $(2n - 3)$-dimensional simplicial complex and a model for $\underline{E\mathrm{Out}(F_n)}$ [118, Section 4.9]. The two results are now available as the preprints [23] and [58].

6.4 ℓ^2-Alexander Torsion

We started off Sect. 6.2 with praising ℓ^2-torsion for being a canonical invariant, independent of any choice of representation as was necessary to define Reidemeister torsion. But not on any account does this mean that there would be nothing to gain if one does decide to consider twisted versions of ℓ^2-torsion. Actually, already introducing a one dimensional twist leads to a surprisingly deep theory on which we shall report in what follows. Towards the end of this section, we will moreover take a quick glance at how these ideas could be further elaborated by considering higher dimensional representations and how they have led to the introduction of *universal ℓ^2-torsion*.

Let X be a connected finite CW complex, set $\pi = \pi_1 X$, and pick some cohomology class

$$\phi \in H^1(X; \mathbb{R}) \cong \mathrm{Hom}(\pi, \mathbb{R}).$$

Every positive real number $t \in (0, \infty)$ defines a ring homomorphism

$$\kappa(\phi, t) \colon \mathbb{Z}\pi \longrightarrow \mathbb{R}\pi, \quad \kappa(\phi, t)(g) = t^{\phi(g)} g$$

by $\mathbb{Z}$-linear extension. We precompose the right $\mathbb{R}\pi$-module structure of $\ell^2\pi$ with $\kappa(\phi, t)$ to construct the $\kappa(\phi, t)$-*twisted ℓ^2-chain complex*

$$C_*^{(2)}(\widetilde{X}; \kappa) = \ell^2\pi \otimes_{\kappa(\phi,t)} C_*(\widetilde{X}).$$

Picking a cellular basis of $\widetilde{X}$ turns $C_*^{(2)}(\widetilde{X}; \kappa)$ again into a chain complex of finitely generated Hilbert modules. So for all $t \in (0, \infty)$ such that $C_*^{(2)}(\widetilde{X}; \kappa(\phi, t))$ is of determinant class, the ℓ^2-torsion is defined according to Definition 6.7. Requiring that in addition $C_*^{(2)}(\widetilde{X}; \kappa(\phi, t))$ be ℓ^2-acyclic (have no reduced homology), we set

$$\tau^{(2)}(X, \phi)(t) = \exp(-\rho^{(2)}(C_*^{(2)}(\widetilde{X}; \kappa))).$$

So we undo taking the logarithm and insert a minus sign that makes sure that determinants are inverted in odd instead of even degree. This convention seems to be customary in the literature on Reidemeister torsion of 3-manifolds. Altering the cellular basis, the base change matrix for $C_n^{(2)}(\widetilde{X}; \kappa)$ will be a generalized permutation matrix with entries $\pm t^{\phi(g_i)}g_i$. Typically, for $t \neq 1$, this matrix will no longer be unitary. The Fuglede–Kadison determinant of such a matrix is $t^{\phi(g_1)+\cdots+\phi(g_{kn})}$ so that the alternating product of determinants, which defines $\tau^{(2)}(X, \phi)(t)$, is only well-defined up to multiplication with a monomial function of the form t^r for some $r \in \mathbb{R}$.

If $D_n \in M(k_n, k_{n-1}; \mathbb{Z}\pi)$ is the matrix representing the n-th cellular differential in $C_*(\widetilde{X})$ with respect to some cellular basis, then $\kappa(\phi, t)(D_n)$, applied entry for entry, is the matrix representing the n-th differential in $C_*^{(2)}(\widetilde{X}; \kappa)$. This implies that if $t \in \mathbb{Q}$ and ϕ lies in the integral lattice $H^1(X; \mathbb{Z}) \subset H^1(X; \mathbb{R})$, then $\kappa(\phi, t)(D_n) \in M(k_n, k_{n-1}; \mathbb{Q}\pi)$ so that $C_*^{(2)}(\widetilde{X}, \kappa(\phi, t))$ is of determinant class if π satisfies the determinant conjecture 5.43. For the moment we artificially set $\tau^{(2)}(X, \phi)(t) = 0$ if either $C_*^{(2)}(\widetilde{X}; \kappa(\phi, t))$ should not be of determinant class or is not ℓ^2-acyclic. In the example of interest, however, Liu showed that this never happens [108].

Theorem 6.17 (Liu [108]) *Suppose N is a connected, compact, irreducible 3-manifold with infinite fundamental group and whose boundary is empty or consists of incompressible tori. Then the function $\tau^{(2)}(N, \phi)$ is continuous and everywhere positive on $(0, \infty)$.*

Of course these assertions do not depend on the particular representative of $\tau^{(2)}(N, \phi)$. The function $\tau^{(2)}(N, \phi)$ is called the *full ℓ^2-Alexander torsion* of N with respect to ϕ. The word "full" is in place because we are working with the universal covering $\widetilde{N}$, which is the "largest" covering of N. Instead, one can also pick some epimorphism $\gamma: \pi \to G$ through which $\phi \in \text{Hom}(\pi, \mathbb{R})$ factorizes and twist the cellular chain complex $C_*^{(2)}(\widetilde{N})$ with

$$\kappa(\phi, \gamma, t)(g) = t^{\phi(g)}\gamma(g).$$

The result is the ℓ^2-*Alexander torsion* function $\tau^{(2)}(N, \gamma, \phi)$ of the regular covering $\overline{N}_{\ker \gamma}$ associated with γ. For example, the abelianization epimorphism $\phi_{ab} \colon \pi \to \mathbb{Z}$ of a knot complement $N = S^3 \setminus \nu K$ gives the ℓ^2-Alexander torsion $\tau^{(2)}(N, \phi_{ab}, \phi_{ab})$ of the canonical infinite cyclic covering $\overline{N}_{[\pi,\pi]}$. To find this function in explicit terms, one can start with a *Wirtinger presentation* $P = \langle g_1, \ldots, g_{k+1} \mid r_1, \ldots, r_k \rangle$ of the knot group π which one can easily read off from a knot diagram as outlined in [46, Section 2]. The corresponding presentation complex X_P is simple homotopy equivalent to N, a folklore result for which a proof is included in [51, Proposition 5.1]. Thus we can replace N by X_P to compute $\tau^{(2)}(N, \phi_{ab}, \phi_{ab})$. The second differential of $C_*(\widetilde{X}_P)$ is realized by the *Fox matrix* $D_2 = (\frac{\partial r_i}{\partial g_j})$ defined in [46, Section 3] and the first differential has the form $D_1 = (g_1 - 1, \ldots, g_{k+1} - 1)$. The abelianization map ϕ_{ab} sends each generator g_i to the generator z of $\langle z \rangle \cong \mathbb{Z}$. It induces the ring homomorphism $\Phi \colon \mathbb{Z}\pi \to \mathbb{Z}[z^{\pm 1}]$ and the matrices $\Phi(D_2)$ and $\Phi(D_1)$ realize the differentials in the chain complex of the infinite cyclic covering of X_P. Deleting any column of $\Phi(D_2)$ gives a square matrix $A_K \in M(k, k; \mathbb{Z}[z^{\pm 1}])$, called the *Alexander matrix* of the knot K. The determinant $\Delta_K = \det_{\mathbb{Z}[z^{\pm 1}]} A_K \in \mathbb{Z}[z^{\pm 1}]$ is called the *Alexander polynomial* of K. Let

$$\Delta_K(z) = a(z - \alpha_1) \cdots (z - \alpha_n)$$

be its complex factorization. Then with the help of *Jensen's formula*, it is not too difficult to see that

$$\tau^{(2)}(N, \phi_{ab}, \phi_{ab})(t) = \max\{t, 1\}^{-1} |a| \prod_{i=1}^{n} \max\{t, |\alpha_i|\}.$$

The factor $\max\{t, 1\}^{-1}$ stems from the first differential. The essential factor

$$M(\Delta_K(tz)) = |a| \prod_{i=1}^{n} \max\{t, |\alpha_i|\}$$

comes from the second differential and is called the *Mahler measure* of the t-scaled Alexander polynomial $\Delta_K(tz)$ of K. This explains why the name "Alexander" shows up in ℓ^2-Alexander torsion. Also recall that the Alexander polynomial Δ_K is only well-defined up to a factor $\pm z^k$. But including such a factor would only multiply the ℓ^2-Alexander torsion with t^k, in beautiful accordance with the flexibility in the definition of $\tau^{(2)}(N, \phi_{ab}, \phi_{ab})$.

Observe that the function $\tau^{(2)}(N, \phi_{ab}, \phi_{ab})$ is *piecewise monomial* and picks up another power of t with each root α_i of Δ_K as soon as $t \geq |\alpha_i|$. As such, the function $\tau = \tau(N, \phi_{ab}, \phi_{ab})$ is *multiplicatively convex* in the sense that for all $t_1, t_2 \in (0, \infty)$ and all $\lambda \in (0, 1)$ we have

$$\tau\left(t_1^{\lambda} \cdot t_2^{1-\lambda}\right) \leq \tau(t_1)^{\lambda} \cdot \tau(t_2)^{1-\lambda}. \tag{6.1}$$

More generally, we obtain a multiplicatively convex function $\tau^{(2)}(N, \gamma, \phi)$ for any 3-manifold N as in Theorem 6.17 whenever $\gamma\colon \pi \to G$ maps onto a virtually abelian group G. One checks this similarly as above, now using higher dimensional Mahler measures if the finite index free abelian subgroup of G has higher rank. It was Liu's clever observation that the property of τ being multiplicatively convex survives when approximating the universal covering by virtually abelian coverings as follows. Since 3-manifold groups are residually finite, we can choose a residual chain (π_i) of π and we consider the characteristic subgroups

$$K_i = \ker(\pi_i \longrightarrow H_1(\pi_i; \mathbb{Q}))$$

of π_i which are normal in π. By construction, the quotient groups $\Gamma_i = \pi/K_i$ are finitely generated virtually abelian. Since the group $(\mathbb{R}, +)$ is torsion-free abelian, any given homomorphism $\phi\colon \pi \to \mathbb{R}$ factorizes through the quotient homomorphisms $\gamma_i\colon \pi \to \Gamma_i$ for all i. Thus we obtain the multiplicatively convex functions $\tau^{(2)}(N, \gamma_i, \phi)$. Liu gives some careful convergence arguments and uses the dangerously subtle continuity properties of the Fuglede–Kadison determinant to conclude the defining inequality (6.1) of multiplicative convexity also for the function $\tau^{(2)}(N, \mathrm{id}_\pi, \phi) \cdot \max\{t, 1\}^m$ with sufficiently large m. Clearly, if a multiplicatively convex function is zero somewhere, it is zero everywhere. But if N_i are the hyperbolic pieces of N, then

$$\tau^{(2)}(N, \phi)(1) = \prod_i \exp\left(\frac{\mathrm{vol}(N_i)}{6\pi}\right) > 0$$

by Example 6.13. So $\tau^{(2)}(N, \phi)$ is positive on $(0, \infty)$. It is also continuous because $\log \circ \tau^{(2)}(N, \phi) \circ \exp$ is convex on $(-\infty, \infty)$ in the ordinary sense and hence, as is well-known, continuous.

This proof method is a lesson for life. Instead of showing a weak property (continuity), one shows a stronger property that includes it (multiplicative convexity), simply because the stronger property is more accessible in the given situation. It's like when you want to steal a car stereo but you can't find the right tool to remove it from the dashboard. Well, it's easier to take the whole car!

If ℓ^2-Alexander torsion were just some continuous function whose value at one gives back a known quantity, it would hardly be worth the trouble. The point is that it carries more interesting geometric information. To explain this, we will go on another quick excursion to 3-manifold theory.

Given a compact oriented surface Σ, possibly disconnected and with boundary, the *complexity* of Σ is defined by $\chi_-(\Sigma) = -\sum_i \chi(\Sigma_i)$ where the sum runs over all those connected components Σ_i of Σ which are *not* homeomorphic to the sphere S^2 or the disk D^2. Every element of $H_2(N, \partial N; \mathbb{Z})$ can be represented by a properly embedded surface $(\Sigma, \partial \Sigma) \subset (N, \partial N)$.

Definition 6.18 The *Thurston norm* of $\phi \in H^1(N; \mathbb{Z})$ is given by

$$x_N(\phi) = \min\{\chi_-(\Sigma) : [\Sigma, \partial\Sigma] \text{ is Poincaré dual to } \phi\}.$$

One can see that $x_N(\phi_1 + \phi_2) \leq x_N(\phi_1) + x_N(\phi_2)$ and $x_N(k\phi) = |k| x_N(\phi)$ for $k \in \mathbb{Z}$, so that x_N extends to a seminorm on $H^1(N; \mathbb{R})$ via the unique extension to $H^1(N; \mathbb{Q})$. This seminorm was first introduced and studied in [166]. The unit ball of x_N is a convex and centrally symmetric polyhedron in the $\mathbb{R}$-vector space $H^1(N; \mathbb{R})$ with finitely many faces each of which lies in a rational affine plane. A priori, the polyhedron can be noncompact because x_N vanishes on the subspace spanned by homologically nontrivial surfaces with nonpositive Euler characteristic. The polyhedron is however known to be compact if N admits a complete hyperbolic metric on the interior.

The geometric significance of this polyhedron is that it allows for a convenient description of all those classes $\phi \in H^1(N; \mathbb{Z}) \cong [N, S^1]$ which in the interpretation as homotopy classes of maps $N \to S^1$ have a representative which is a surface bundle over the circle. For such *fibered class* $N \to S^1$, any fiber $(\Sigma, \partial\Sigma) \subset (N, \partial N)$ is Poincaré dual to ϕ and norm realizing, meaning $x_N(\phi) = \chi_-(\Sigma)$. Thurston showed that the fibered classes are precisely the integral points in the open cones over certain top dimensional faces in the polyhedron, the so-called *fibered cones* lying over *fibered faces*. It can happen that the x_N-unit ball has no fibered faces at all. But Agol [3, Theorem 5.1] showed that given a non-trivial, non-fibered class $\phi \in H^1(N, \mathbb{Z})$, there exists a finite covering $p : \overline{N} \to N$ such that $p^*\phi$ lies in the boundary of a fibered cone, provided $\pi = \pi_1 N$ is *RFRS*. This acronym is short for *residually finite rationally solvable* and means that π has a residual chain (π_i) such that each map $\pi_i \to \pi_i/\pi_{i+1}$ factors through $\pi_i \to H_1(\pi_i)_{\text{free}}$. Moreover, if π is infinite RFRS, then N has a finite covering with positive first Betti number. So on this covering, one can pick a nontrivial class and if it is not already fibered, then by the above, yet another finite covering has a unit Thurston norm ball with a fibered face. Soon thereafter, Agol [4] and Wise [174] showed that $\pi_1 N$ is virtually RFRS if N is hyperbolic and Przytycki and Wise [148] extended this result to the case where N has a hyperbolic piece, in other words is not a graph manifold. So these manifolds N always have a finite covering with fibered faces in their polytopes. In particular, this settles the famous *virtually fibered conjecture*.

Theorem 6.19 (Virtually Fibered Theorem) *Suppose N is a connected, compact, irreducible 3-manifold with infinite fundamental group and whose boundary is empty or consists of incompressible tori. If N is not a graph manifold, then some finite covering of N is a surface bundle over the circle.*

As an alternative to reading Agol's original proof [3, Theorem 5.1], the reader can also find a beautiful treatment in [48]. A thoroughly attributed exposition of how these results fit into the web of all the spectacular recent breakthroughs in 3-manifold theory is given in [7, Chapters 4 and 5].

We got somewhat carried away but you should be convinced by now that the Thurston norm is a central tool in the study of 3-manifolds. Thus it is a proud feature of the full ℓ^2-Alexander torsion that it recovers the Thurston norm. To see this, Liu [108, Theorem 1.2.2] proved that $\tau^{(2)}(N, \phi)$ is *asymptotically monomial* which implies that the limits

$$d_\infty = \lim_{t \to \infty} \frac{\log \tau^{(2)}(N, \phi)(t)}{\log t} \quad \text{and} \quad d_0 = \lim_{t \to 0^+} \frac{\log \tau^{(2)}(N, \phi)(t)}{\log t}$$

exist. The real number $\deg \tau^{(2)}(N, \phi) = d_\infty - d_0$ is called the *asymptotic degree* of $\tau^{(2)}(N, \phi)$. We remark that by Dubois et al. [39], ℓ^2-Alexander torsions are *symmetric* in general, meaning $\tau^{(2)}(N, \gamma, \phi)(t^{-1}) = t^r \tau^{(2)}(N, \gamma, \phi)(t)$ for some $r \in \mathbb{R}$, so clearly one of the two limits above exists if and only if the other does.

Theorem 6.20 *Suppose N is a connected, compact, irreducible 3-manifold with infinite fundamental group and whose boundary is empty or consists of incompressible tori. Then for all $\phi \in H^1(N; \mathbb{R})$, we have*

$$\deg \tau^{(2)}(N, \phi) = x_N(\phi).$$

The result is likewise due to Liu [108, Theorem 1.2.3] and independently to Friedl and Lück [49] (for $\phi \in H^1(N; \mathbb{Q})$). Moreover, the theorem had a precursor for the (p, q)-torus knot complement $N_{p,q}$ in which case the earlier defined ℓ^2-*Alexander invariant* of Li and Zhang [104] was computed by Dubois and Wegner [38] in terms of the knot genus $g = (p - 1)(q - 1)/2$. With our notation, they showed that

$$\tau^{(2)}(N_{p,q}, \phi_{ab})(t) = \max\{1, t\}^{pq-p-q} = \max\{1, t\}^{2g-1}.$$

Note that such a simple formula is only possible because torus knots are not hyperbolic as reflected in the property $\tau^{(2)}(N_{p,q}, \phi)(1) = 1$. The formula accords with Theorem 6.20 because Seifert surfaces are always dual to ϕ_{ab} so that $x_N(\phi_{ab}) \le 2g - 1$ for any nontrivial knot, and this equality is in fact an equality, see for example [47, Lemma 2.2]. Building on work of Herrmann [72], Dubois et al. [40, Theorem 1.2] generalized the torus knot computation to

$$\tau^{(2)}(N, \phi)(t) = \max\{1, t\}^{x_N(\phi)}$$

if $N \ne D^2 \times S^1, S^2 \times S^1$ is a graph manifold and $\phi \in H^1(N; \mathbb{R})$ is any nontrivial class. Together with Lück–Schick's result that $\tau^{(2)}(N, \phi_{ab})(1) > 1$ for non-graph manifolds, this implies that $\tau^{(2)}(N, \phi_{ab}) = \max\{t, 1\}^{-1}$ if and only if K is the unknot. In other words, Zhang-Li's ℓ^2-Alexander invariant detects the unknot, a fact first noticed by Ben Aribi [15].

We conclude this section by drawing the reader's attention to two new research directions emerging out of the above. Firstly, we can take the viewpoint that given

$\phi\colon \pi \to \mathbb{R}$, the elements $t \in (0, \infty)$ parametrize the family of one-dimensional $\mathbb{R}$-representations of π given by multiplication with t^ϕ. In that sense, the full ℓ^2-Alexander torsion $\tau^{(2)}(N, \phi)$ is merely a baby example of the idea to consider ℓ^2-torsion as a function on *representation varieties*. The setup for this idea would be roughly as follows. We fix an ℓ^2-acyclic free finite G-CW complex X of determinant class, and consider varying, say complex, finite dimensional G-representations V. Then we first form $V \otimes_{\mathbb{C}} C_*(X; \mathbb{C})$, afterwards we pass to the ℓ^2-completion by applying $\ell^2 G \otimes_{\mathbb{C}G}$ with respect to the diagonal action, and finally we take the ℓ^2-torsion. Well-definedness questions about ℓ^2-acyclicity and determinant class already require quite some effort. Continuity questions—from today's point of view—seem to be almost out of reach. To some extent, this is also related to the hard convergence questions of ℓ^2-torsion we will consider in the next section. But this is also why any minuscule progress here could be valuable. Lück has launched a first attack on these questions and the reader can find out about it in the technical paper [119].

The second outcome of ℓ^2-Alexander torsion arises after realizing that the basic properties of ordinary ℓ^2-torsion as listed in Theorem 6.9 can be reproven for twisted versions like the full ℓ^2-Alexander torsion with virtually unchanged arguments, as is for instance done in [38, Proposition 2.23]. This indicates that the properties are true in a *universal* sense and should be proven once and for all on a more abstract level. In the concrete case of ℓ^2-torsion, one achieves this by not taking the Fuglede–Kadison determinant too early but instead considering the weak isomorphism between odd and even Hilbert chain modules of an ℓ^2-acyclic chain complex as an element in the first *weak algebraic K-theory* $K_1^w(\mathbb{Z}G)$ of the ring $\mathbb{Z}G$. Similarly as ordinary first algebraic K-theory, $K_1^w(\mathbb{Z}G)$ has endomorphisms $(\mathbb{Z}G)^n \to (\mathbb{Z}G)^n$ as generators though these are not required to be $\mathbb{Z}G$-isomorphism but only weak isomorphisms after ℓ^2-completion. Relations are likewise defined in terms of weak isomorphisms instead of $\mathbb{Z}G$-isomorphisms. For an ℓ^2-acyclic finite free G-CW complex X, *universal ℓ^2-torsion* $\rho_u(X)$ then lies in the quotient $\mathrm{Wh}^w(G)$ of $K_1^w(\mathbb{Z}G)$ called the *weak Whitehead group*, obtained by factoring out trivial units so that $\rho_u(X)$ is a well-defined invariant, independent of a choice of cellular basis. Many familiar properties of ℓ^2-torsion can already be proven for universal ℓ^2-torsion so that they are right away available for images of $\rho_u(X)$ like ordinary ℓ^2-torsion or ℓ^2-Alexander torsion. Also interesting is the *polytope homomorphism*

$$\mathbb{P}\colon \mathrm{Wh}^w(G) \longrightarrow \mathcal{P}_{\mathbb{Z}}^{\mathrm{Wh}}(H_1(G)_{\mathrm{free}})$$

to the Grothendieck completion of integral polytopes in $H_1(G)_{\mathrm{free}} \otimes_{\mathbb{Z}} \mathbb{R}$ with addition given by Minkowski sum up to integral translation. For a 3-manifold N as in the above theorems and assuming the Atiyah conjecture for $\pi_1 N$, it turns out that $2\mathbb{P}(\rho_u(\widetilde{N}))$ is dual to the Thurston polytope. The reader interested in this new approach and the mentioned applications is directed to Friedl and Lück [50].

6.5 Torsion in Homology

As announced at the end of Sect. 6.2, we will now discuss another striking parallelism between ℓ^2-torsion and Euler characteristic. It occurs in a field that has attracted massive research effort in recent years: *homology growth*. Lück's approximation theorem can be seen as a fundamental result in this area. If X is a connected finite CW complex with fundamental group $G = \pi_1 X$, then a positive n-th ℓ^2-Betti number $b_n^{(2)}(\widetilde{X}) > 0$ detects *linear free homology growth* in degree n. This means that along the coverings $\overline{X}_i$ of X corresponding to any residual chain (G_i) in G, the rank of the free part of $H_n(X_i)$ grows *asymptotically proportionally* to the index $[G : G_i]$. The asymptotic proportionality constant is precisely $b_n^{(2)}(\widetilde{X})$. In this vein, the Singer conjecture (Conjecture 1.5) predicts the following phenomenon for even dimensional *aspherical* manifolds.

Conjecture 6.21 Let X be an aspherical, $2n$-dimensional, closed, connected manifold with residually finite fundamental group $G = \pi_1 X$. Then for every residual chain (G_i) in G we have

$$\lim_{i \to \infty} \frac{\operatorname{rank}_{\mathbb{Z}} H_n(\overline{X}_i)_{\text{free}}}{[G : G_i]} = (-1)^n \chi(X).$$

The left hand side equals $b_n^{(2)}(\widetilde{X})$ by Lück's approximation theorem and the right hand side equals $b_n^{(2)}(\widetilde{X})$ if the Singer conjecture holds true. So the Singer conjecture says that a non-zero Euler characteristic detects free homology growth in middle degree for even dimensional aspherical manifolds. Here is the odd dimensional cousin of this conjecture.

Conjecture 6.22 Let X be an aspherical, $(2n + 1)$-dimensional, closed, connected manifold with residually finite fundamental group $G = \pi_1 X$. Then for every residual chain (G_i) in G we have

$$\lim_{i \to \infty} \frac{\log |H_n(\overline{X}_i)_{\text{tors}}|}{[G : G_i]} = (-1)^n \rho^{(2)}(\widetilde{X}).$$

So conjecturally, non-zero ℓ^2-torsion detects exponential growth of torsion in middle degree homology of an odd-dimensional aspherical manifold. Be aware that the conjecture also incorporates the Singer conjecture in the sense that an odd-dimensional, aspherical manifolds should be ℓ^2-acyclic. By Corollary 5.50, the manifold X is moreover of determinant class. Here and elsewhere we assume that X is endowed with some CW structure. Such a structure always exists for smooth manifolds. For topological manifolds it exists except possibly in dimension four which is irrelevant to Conjecture 6.22.

To understand the philosophy behind Conjecture 6.22 we introduce yet another torsion invariant. It is known as *integral torsion*, sometimes also *Milnor torsion*, and builds the bridge from ℓ^2-torsion to torsion in homology.

Definition 6.23 Let X be a finite (non-equivariant) CW complex. Then the *integral torsion* of X is given by

$$\rho^{\mathbb{Z}}(X) = \sum_{n \geq 0} (-1)^n \log |H_n(X)_{\text{tors}}|.$$

Of course we can consider a non-equivariant CW complex X as a G-CW complex X for the trivial group $G = \{1\}$. Let us compare integral torsion to ℓ^2-torsion in this case. Strictly speaking, the ℓ^2-torsion of X is not even defined because X has at least a positive zeroth ℓ^2-Betti number. Nevertheless, we will use the notation $\rho^{(2)}(\{1\} \curvearrowright X)$ or just $\rho^{(2)}(X)$ which are to be understood as $\rho^{(2)}(C_*^{(2)}(X))$ in the sense of Definition 6.7. We only need to keep in mind that $\rho^{(2)}(X)$ now depends on the specific CW structure of X. After an index shift in the defining sum of $\rho^{(2)}(X)$, we obtain

$$\rho^{\mathbb{Z}}(X) - \rho^{(2)}(X) = \sum_{n \geq 0} (-1)^n \log \frac{|H_n(X)_{\text{tors}}|}{\det_{\mathcal{R}\{1\}} d_{n+1}^{(2)}}.$$

The torsion group order $|H_n(X)_{\text{tors}}|$ is given by the absolute value of the product of the nonzero *invariant factors* of the $\mathbb{Z}$-module homomorphism d_{n+1}, see for example [87, Lemma 6]. The Fuglede–Kadison determinant $\det_{\mathcal{R}\{1\}} d_{n+1}^{(2)}$, in turn, is given by the product of the positive *singular values* of the operator

$$d_{n+1}^{(2)} : C_{n+1}(X; \mathbb{C}) \longrightarrow C_n(X; \mathbb{C})$$

as we explained below Definition 5.42. (Also recall from Example 3.14 that $C_n^{(2)}(X) = C_n(X; \mathbb{C})$ because G is trivial.) We remind the reader that the singular values of an operator A of finite dimensional Hilbert spaces are by definition the eigenvalues of the operator $|A| = \sqrt{A^*A}$. Thus if the differentials in the cellular chain complex happen to be diagonal matrices with respect to some fixed cellular basis (meaning the (i, j)-th entry can be nonzero only if $i = j$), then invariant factors and singular values coincide and ℓ^2-torsion equals integral torsion. In general, however, the two concepts are distinct and so called *regulators* identify the difference. Let $H_n(X)_{\text{free}} = H_n(X)/H_n(X)_{\text{tors}}$ be the free part of the n-th homology. As "$\mathbb{C}$ is flat over $\mathbb{Z}$", we have a canonical isomorphism $\alpha_n \colon \mathbb{C} \otimes_{\mathbb{Z}} H_n(X)_{\text{free}} \xrightarrow{\sim} H_n(C_*(X; \mathbb{C}))$.

Definition 6.24 Pick any $\mathbb{Z}$-basis of $H_n(X)_{\text{free}}$ to endow $\mathbb{C} \otimes_{\mathbb{Z}} H_n(X)_{\text{free}}$ with an inner product. Then the n-th *regulator* of X is given by

$$R_n(X) = \log \det_{\mathcal{R}\{1\}} \alpha_n.$$

If we change the $\mathbb{Z}$-basis of $H_n(X)_{\text{free}}$, then α_n gets multiplied by the transition matrix which is invertible over $\mathbb{Z}$. It thus has determinant ± 1, so that the Fuglede–Kadison determinant remains unchanged.

Theorem 6.25 *Let X be a finite CW complex. Then*

$$\rho^{\mathbb{Z}}(X) - \rho^{(2)}(X) = \sum_{n \geq 0} (-1)^n R_n(X).$$

Proof We describe a procedure to construct $\mathbb{Z}$-bases of the cellular chain groups of X which diagonalize all differentials. To begin with, let us introduce the standard notation $C_n = C_n(X)$ for the chain groups, $Z_n = \ker d_n$ for the n-cycles and $B_n = \operatorname{im} d_{n+1}$ for the n-boundaries. Let T_n be the kernel of the canonical homomorphism $p_n \colon Z_n \to H_n(X)_{\text{free}}$. The image $\operatorname{im} p_n$ is a submodule of $H_n(X)_{\text{free}}$, hence free, so we can lift it to a submodule $H_n \subseteq Z_n$. Similarly, the image of $d_n \colon C_n \to C_{n-1}$ is free and we pick a lift $S_n \subseteq C_n$. Since $B_n \subseteq T_n$, the differential d_{n+1} restricts to a homomorphism $S_{n+1} \to T_n$ of free $\mathbb{Z}$-modules which moreover has finite cokernel. Thus S_{n+1} and T_n have equal rank. Pick bases of S_{n+1} and T_n with respect to which the homomorphism has Smith normal form. It is thus given by a diagonal matrix D_{n+1} with entries the nonzero invariant factors of d_{n+1}. Finally, pick any $\mathbb{Z}$-basis of H_n. We have constructed direct sum decompositions which we agree to order as

$$C_{2n+1} = S_{2n+1} \oplus H_{2n+1} \oplus T_{2n+1} \quad \text{and} \quad C_{2n} = T_{2n} \oplus H_{2n} \oplus S_{2n}$$

for the odd and even chain groups. This effects that the differentials d_{2n+1} and d_{2n} have the block form

$$\begin{pmatrix} D_{2n+1} & 0 & 0 \\ 0 & 0 & 0 \\ 0 & 0 & 0 \end{pmatrix} \quad \text{and} \quad \begin{pmatrix} 0 & 0 & 0 \\ 0 & 0 & 0 \\ 0 & 0 & D_{2n} \end{pmatrix}.$$

with respect to the constructed basis of the chain complex. It makes sense to call such a basis *differentially adapted*. Let us complexify our free abelian groups S_*, H_* and T_* to $\mathbb{C}$-vector spaces by applying the functor $(\cdot)^{\mathbb{C}} = \mathbb{C} \otimes_{\mathbb{Z}} (\cdot)$. We obtain an isomorphism

$$D \colon C_{2*+1}^{\mathbb{C}} \oplus H_{2*}^{\mathbb{C}} \xrightarrow{\cong} C_{2*}^{\mathbb{C}} \oplus H_{2*+1}^{\mathbb{C}}$$

where in this context the symbol "$*$" means direct sum over all $* = n$. To wit, with respect to the above decompositions of C_* and the chosen bases, the isomorphism D is implemented by the invertible (4×4)-block matrix

$$\begin{pmatrix} D_{2*+1} & 0 & 0 & 0 \\ 0 & 0 & 0 & 1 \\ 0 & 0 & D_{2*}^{-1} & 0 \\ 0 & 1 & 0 & 0 \end{pmatrix}.$$

We endow the $\mathbb{C}$-vector spaces $C^{\mathbb{C}}_{2*+1} \oplus H^{\mathbb{C}}_{2*}$ and $C^{\mathbb{C}}_{2*} \oplus H^{\mathbb{C}}_{2*+1}$ with the inner products for which the canonically included differentially adapted basis is orthonormal. Recall that the ℓ^2-chain complex $C^{(2)}_*(X) = C_*(X; \mathbb{C})$ is likewise endowed with inner products and for these inner products any cellular basis is orthonormal. Accordingly, the filtration

$$0 \subseteq \operatorname{im} d^{(2)}_{n+1} \subseteq \ker d^{(2)}_n \subseteq C^{(2)}_n(X)$$

by subspaces determines orthogonal decompositions

$$C^{(2)}_{2*+1}(X) \cong (\ker d^{(2)}_{2*+1})^\perp \oplus H^{(2)}_{2*+1}(X) \oplus \operatorname{im} d^{(2)}_{2*+2},$$

$$C^{(2)}_{2*}(X) \cong \operatorname{im} d^{(2)}_{2*+1} \oplus H^{(2)}_{2*}(X) \oplus (\ker d^{(2)}_{2*})^\perp$$

where $H^{(2)}_*(X)$ sits in $C^{(2)}_*(X)$ as the orthogonal complement of $\operatorname{im} d^{(2)}_{n+1}$ in $\ker d^{(2)}_n$. Similarly as above, we obtain an isomorphism

$$D^{(2)} : C^{(2)}_{2*+1}(X) \oplus H^{(2)}_{2*}(X) \xrightarrow{\cong} C^{(2)}_{2*}(X) \oplus H^{(2)}_{2*+1}$$

which has the orthogonal block decomposition

$$\begin{pmatrix} d^{(2)\perp}_{2*+1} & 0 & 0 & 0 \\ 0 & 0 & 0 & 1 \\ 0 & 0 & \left(d^{(2)\perp}_{2*}\right)^{-1} & 0 \\ 0 & 1 & 0 & 0 \end{pmatrix}.$$

Here $d^{(2)\perp}_* : (\ker d^{(2)}_*)^\perp \longrightarrow \operatorname{im} d^{(2)}_*$ is the isomorphism induced by $d^{(2)}_*$. Let $s_* : C^{(2)}_*(X) \to C^{(2)}_*(X)$ be the composition of the following two *shear transformations*: The first leaves invariant the subspace $\operatorname{im} d^{(2)}_{*+1} \oplus (\ker d^{(2)}_*)^\perp$ and restricts on $H^{\mathbb{C}}_*$ to the orthogonal projection onto $H^{(2)}_*(X)$. The second leaves invariant the subspace $\operatorname{im} d^{(2)}_{*+1} \oplus H^{(2)}_*(X)$ and projects the image of $S^{\mathbb{C}}_*$ under the first transformation orthogonally to $(\ker d^{(2)}_*)^\perp$. We obtain a commutative diagram of Hilbert $\mathcal{L}\{1\}$-module isomorphism

$$
\begin{array}{ccccc}
C^{\mathbb{C}}_{2*+1} \oplus H^{\mathbb{C}}_{2*} & \xrightarrow{\operatorname{id} \oplus p^{\mathbb{C}}_{2*}} & C^{(2)}_{2*+1}(X) \oplus H^{\mathbb{C}}_{2*}(X)_{\mathrm{free}} & \xrightarrow{s_{2*+1} \oplus \alpha_{2*}} & C^{(2)}_{2*+1}(X) \oplus H^{(2)}_{2*}(X) \\
\Big\downarrow{\scriptstyle D} & & & & \Big\downarrow{\scriptstyle D^{(2)}} \\
C^{\mathbb{C}}_{2*} \oplus H^{\mathbb{C}}_{2*+1} & \xrightarrow{\operatorname{id} \oplus p^{\mathbb{C}}_{2*+1}} & C^{(2)}_{2*} \oplus H^{\mathbb{C}}_{2*+1}(X)_{\mathrm{free}} & \xrightarrow{s_{2*} \oplus \alpha_{2*+1}} & C^{(2)}_{2*}(X) \oplus H^{(2)}_{2*+1}(X)
\end{array}
$$

where α_n is the isomorphism in Definition 6.24. Beware that the identity map $C_*^{\mathbb{C}} \xrightarrow{\text{id}} C_*^{(2)}(X)$ is typically not an isometry with respect to the inner products we assigned to $C_*^{\mathbb{C}}$ and $C_*^{(2)}(X)$. However, the transition matrix from the differentially adapted basis to the cellular basis is invertible over $\mathbb{Z}$, thus has determinant ± 1. It follows that $\det_{\mathcal{R}\{1\}} \text{id} = 1$. We can endow $H_*(X)_{\text{free}}$ with the image of our $\mathbb{Z}$-basis of H_* under p_* so that $p_*^{\mathbb{C}}$ becomes a unitary, hence also $\det_{\mathcal{R}\{1\}} p_*^{\mathbb{C}} = 1$. Shear transformations have block diagonal form $\left(\begin{smallmatrix} 1 & * \\ 0 & 1 \end{smallmatrix}\right)$, so $\det_{\mathcal{R}\{1\}} s_* = 1$. The spectral measures of $|d_*^{(2)}|$ and $|d_*^{(2)\,\perp}|$ only differ at the point zero which implies $\det_{\mathcal{R}\{1\}} d_*^{(2)} = \det_{\mathcal{R}\{1\}} d_*^{(2)\,\perp}$. Finally, since $\det = \det_{\mathcal{R}\{1\}}$ is multiplicative on compositions of isomorphisms, the above diagram gives

$$\prod_{m\geq 0} \det \alpha_{2m} \cdot \prod_{n\geq 0} \det d_n^{(2)\,(-1)^{n+1}} = \prod_{n\geq 0} |H_n(X)_{\text{tors}}|^{(-1)^n} \cdot \prod_{m\geq 0} \det \alpha_{2m+1}.$$

Note moreover that $\det_{\mathcal{R}\{1\}} d_0^{(2)} = \det_{\mathcal{R}\{1\}} 0 = 1$, so we can leave out the zeroth factor in the second product. Taking log completes the proof.   □

It is instructive to illustrate the vertical isomorphisms from the commutative diagram appearing in the proof by our beer coaster picture. Since we are working with finite CW complexes and trivial coefficients, the cellular chain complex will always have nontrivial homology. This means we have gaps between our beer coasters which prevent the odd part $C_{2*+1}^{\mathbb{C}}$ from being isomorphic to the even part $C_{2*}^{\mathbb{C}}$. However, we want them to be isomorphic because we know the Fuglede–Kadison determinant is multiplicative for compositions of isomorphisms. So the pragmatic solution is to "fill the gaps" between the beer coasters, and add the even homology $H_{2*}^{\mathbb{C}}$ to $C_{2*+1}^{\mathbb{C}}$ and the odd homology $H_{2*+1}^{\mathbb{C}}$ to $C_{2*}^{\mathbb{C}}$. We do the same thing for the other vertical isomorphism and fill the gaps in $C_{2*+1}^{(2)}(X)$ and $C_{2*}^{(2)}(X)$ with the ℓ^2-homology. Finally, the horizontal maps identify the two isomorphisms and only the regulators α_{2*} and α_{2*+1} have non-unital Fuglede–Kadison determinant. This gives the asserted formula.

Finally, we have collected all the preliminaries and are in a position to outline a tentative proof strategy for Conjecture 6.22.

"Proof" of Conjecture 6.22 The first ingredient we would need is a proof of a *Singer conjecture* for torsion in homology that would assert

$$(-1)^n \frac{\log |H_n(\overline{X}_i)_{\text{tors}}|}{[G : G_i]} \approx \frac{\rho^{\mathbb{Z}}(\overline{X}_i)}{[G : G_i]}$$

for large i. In words, torsion in homology should asymptotically be concentrated in the middle degree so that all but the middle summand in the alternating sum $\rho^{\mathbb{Z}}$ can be neglected. In an arithmetic setting, this is also suspected to be true by the so-called *Bergeron–Venkatesh conjecture* which we will present in the next section. The second ingredient would be a proof of a *small regulators conjecture* that should say that the alternating sum $\sum_{n \geq 0} (-1)^n R_n(\overline{X}_i)$ divided by $[G : G_i]$ should become small for large i. Then Theorem 6.25 would give

$$\frac{\rho^{\mathbb{Z}}(\overline{X}_i)}{[G : G_i]} \approx \frac{\rho^{(2)}(\{1\} \curvearrowright \overline{X}_i)}{[G : G_i]}$$

for large i. Considering $\overline{X}_i$ as a non-equivariant CW complex implicates again that $\rho^{(2)}(\{1\} \curvearrowright \overline{X}_i)$ depends a priori on the CW structure of $\overline{X}_i$ because $\overline{X}_i$ has nontrivial zeroth homology. Similar remarks apply if we consider $\overline{X}_i$ as a G/G_i-CW complex. As such it gives rise to a chain complex of Hilbert $\mathcal{L}(G/G_i)$-modules and

$$\frac{\rho^{(2)}(\{1\} \curvearrowright \overline{X}_i)}{[G : G_i]} = \rho^{(2)}(G/G_i \curvearrowright \overline{X}_i)$$

because we observed below Definition 5.42 that for a finite group H, the Fuglede–Kadison determinant is the $|H|$-th root of the product of positive singular values. The third and final ingredient to the proof is the *determinant approximation conjecture* stated in Remark 5.48. If true, it would immediately allow the conclusion

$$\lim_{i \to \infty} \rho^{(2)}(G/G_i \curvearrowright \overline{X}_i) = \rho^{(2)}(\widetilde{X}). \qquad \square$$

Each of the three ingredients, the Singer conjecture for torsion, the small regulator conjecture, and the determinant approximation conjecture is a huge problem by itself; and each is of independent interest. At the time of writing, all of them are wide open. Let us however take this opportunity to discuss a possible proof strategy for the determinant approximation conjecture 5.48 suggested by Lück [120, Section 16]. The determinant conjecture 5.43 implies the logarithmic estimate (5.10). The determinant approximation conjecture 5.48 would follow, if we could improve this estimate as specified in the following theorem.

Theorem 6.26 *Suppose that for a given residual system $(G_i)_{i \in I}$ and $A \in M(k, l; \mathbb{Q}G)$, there exist constants $C, \delta > 0$ and $0 < \varepsilon < 1$ such that*

$$\mu_{|\cdot A_i|}((0, \lambda)) \leq \frac{C}{|\log \lambda|^{1+\delta}}$$

for all $i \in I$ and all $\lambda < \varepsilon$. Then the determinant approximation conjecture 5.48 is true for G, (G_i), and A.

Proof The inequality $\det_{\mathcal{R}(G)} \cdot A \geq \limsup_{i \in I} \det_{\mathcal{R}(G/G_i)} \cdot A_i$ follows exactly as in Proposition 5.47 because we are assuming an even sharper bound as the one in (5.10).

Now again, let $\mu_i = \mu_{|\cdot A_i|}$. Differentiation gives the density appearing in the measure

$$\frac{C(\delta+1)}{x \cdot (-\log(x))^{\delta+2}} \, dx$$

whose value on $(0, \lambda)$ is the logarithmic bound $C/(-\log \lambda)^{1+\delta}$. Since the logarithm is a monotone increasing negative function on $(0, \varepsilon)$, it follows that for all $\lambda < \varepsilon$ and all $i \in I$ we have

$$\int_{0+}^{\lambda^-} \log d\mu_i \geq -C(\delta+1) \int_{0+}^{\lambda} \frac{1}{x(-\log(x))^{\delta+1}} \, dx = -\frac{C(\delta+1)}{\delta(-\log(\lambda))^{\delta}}.$$

By the usual argument from Proposition 5.16, we know that the measures μ_i converge weakly to $\mu = \mu_{|\cdot A|}$ on every compact interval $[\lambda, a]$ with $0 < \lambda < \varepsilon$ and $a = k \cdot \sqrt{\|A^*A\|_1}$. Thus we obtain

$$\liminf_{i \in I} \int_{0+}^{a} \log d\mu_i \geq -\frac{C(\delta+1)}{\delta(-\log(\lambda))^{\delta}} + \int_{\lambda}^{a} \log d\mu$$

for all $0 < \lambda < \varepsilon$, hence also

$$\liminf_{i \in I} \int_{0+}^{a} \log d\mu_i \geq \int_{0+}^{a} \log d\mu.$$

Applying the exponential function to this inequality gives

$$\det\nolimits_{\mathcal{R}(G)} \cdot A \leq \liminf_{i \in I} \det\nolimits_{\mathcal{R}(G/G_i)} \cdot A_i. \qquad \qquad \square$$

The sharpening of the logarithmic bound (5.10) demanded in Theorem 6.26 might look innocuous but rest assured it is not. In fact, Grabowski [60] constructs for each $\delta > 0$ a group G_δ and a self-adjoint $S_\delta \in \mathbb{Z}G_\delta$ such that

$$\mu_{S_\delta}((0, \lambda)) > \frac{C}{|\log \lambda|^{1+\delta}}$$

for some constant $C > 0$ and all small $\lambda > 0$. The groups G_δ are wreath products of similar type as were used to answer Atiyah's Question 3.27. Grabowski's result shows that the order of quantors in Theorem 6.26 is important. The constants will have to depend at least on G. For more information on approximation questions, including the relation to approximating *analytic* ℓ^2-torsion, the reader is referred to the survey article [120].

6.6 Torsion in Twisted Homology

In Sect. 6.5 we saw that Conjecture 6.22 expects that ℓ^2-torsion should detect exponential torsion growth in middle degree homology of an odd dimensional aspherical manifold X. The homology groups of interest were the torsion subgroups of $H_n(\overline{X}_i) = H_n(\overline{X}_i; \mathbb{Z})$ for a residual tower of finite Galois coverings $\overline{X}_i$ of X if X is $(2n + 1)$-dimensional. Of course integer coefficients are the canonical choice to work with and it certainly does not make sense to consider coefficients in representations V of $\pi_1 X$ over fields, as we did in Sect. 6.4, because homology would then consist of vector spaces so that there is no torsion left to investigate. But often the manifold of interest X arises from a certain geometric context in which the fundamental group *stabilizes some $\mathbb{Z}$-lattice $M \subset V$*, so that M is the $\mathbb{Z}$-span of some basis in the vector space V over a field of characteristic zero and M is an invariant subset of the $\pi_1 X$-action. In that case, the free abelian group M is turned into a finitely generated $\mathbb{Z}(\pi_1 X)$-module and it is meaningful to ask for the amount and growth of the torsion subgroup of $H_n(\overline{X}_i; M)$.

Such contexts are typical for *arithmetic groups* on which we shall now spend a page or so for a fusillade of definitions and facts. A *linear algebraic group* $\mathbf{G}$ defined over $\mathbb{Q}$ is a subgroup of $\mathrm{GL}(n; \mathbb{C})$ which is *Zariski closed* over $\mathbb{Q}$, meaning it is the zero locus of a set of polynomials in the n^2 matrix entries with coefficients in $\mathbb{Q}$. An example would be $\mathbf{G} = \mathbf{SL_n}$ which is defined by the polynomial $p(A_{ij}) = \det(A_{ij}) - 1$. We set $\mathbf{G}(\mathbb{R}) = \mathbf{G} \cap \mathrm{GL}(n; \mathbb{R})$ and similarly $\mathbf{G}(\mathbb{Z}) = \mathbf{G} \cap \mathrm{GL}(n; \mathbb{Z})$. We say that a subgroup $\Gamma \subset \mathbf{G}$ is called *arithmetic* if it is *commensurable* with $\mathbf{G}(\mathbb{Z})$, so that the intersection $\Gamma \cap \mathbf{G}(\mathbb{Z})$ has finite index both in Γ and in $\mathbf{G}(\mathbb{Z})$. Any element of $\mathbf{G}(\mathbb{Z})$ survives in the finite quotient group $\mathbf{G}(\mathbb{Z}/n)$ obtained by reducing matrix coefficients mod n for big enough n. This shows that arithmetic groups are residually finite. We say that an arithmetic subgroup of $\mathbf{G}$ is a *congruence subgroup* if for some n it contains the kernel of $\mathbf{G}(\mathbb{Z}) \to \mathbf{G}(\mathbb{Z}/n)$ as a finite index subgroup. The kernels themselves are termed *principal congruence subgroups*. Already Felix Klein knew that many (in fact most) arithmetic subgroups of $\mathbf{SL_2}$ are not congruence subgroups. In contrast, for $n \geq 3$ all arithmetic subgroups of $\mathbf{SL_n}$ are congruence subgroups [11]. A little less restrictively, we will say that $\mathbf{G}$ satisfies the *congruence subgroup property* or, for short, "$\mathbf{G}$ has CSP" if $\mathbf{G}(\mathbb{Z})$ has a finite index subgroup Γ such that all finite index subgroups of Γ are congruence subgroups. For a quick overview on the congruence subgroup property, the reader may consult [89, Sections 2.2 and 2.3]. For an extensive survey, we recommend [147]

We say that $\mathbf{G}$ is *semisimple* if the trivial group is the only connected solvable normal subgroup of $\mathbf{G}$. For example $\mathbf{SL_n}$ is semisimple whereas $\mathbf{GL_n}$ is not because it has center isomorphic to $\mathbb{C}^*$ given by constant diagonal matrices. Note that treating $\mathbf{GL_n}$ as a linear algebraic group needs proof. It embeds into $\mathrm{GL}(n + 1; \mathbb{C})$ via $g \mapsto \begin{pmatrix} g & 0 \\ 0 & (\det g)^{-1} \end{pmatrix}$ and the image is defined by polynomial equations: with the exception of the lower right corner entry $x_{n+1,n+1}$, all entries in the last row and column are required to vanish and in addition we require $x_{n+1,n+1} \cdot \det(x_{ij}) - 1 = 0$ where x_{ij} is the matrix with the last column and row deleted. The product of two linear algebraic groups $\mathbf{G_1}$ and $\mathbf{G_2}$ is linear algebraic as one can see by using a block diagonal embedding into $GL(n_1+n_2; \mathbb{C})$. In particular, $(\mathbb{C}^*)^n = \mathbf{GL_1} \times \cdots \times \mathbf{GL_1} = \mathbf{T}^n$ is a linear algebraic group defined over $\mathbb{Q}$ to which we want to refer as the *standard n-dimensional torus*. A group homomorphism $\mathbf{G_1} \to \mathbf{G_2}$ is called a *k-morphism* for some field $\mathbb{Q} \subseteq k \subseteq \mathbb{C}$ if after embedding $\mathbf{G_i}$ via $\mathbf{GL_n}$ into $GL(n_i + 1; \mathbb{C})$ as above, the entries of $\mathbf{G_2}$ are polynomials in the entries of $\mathbf{G_1}$ with coefficients in k. A linear algebraic $\mathbb{Q}$-group $\mathbf{S}$ is called an *n-dimensional torus* if it is $\mathbb{C}$-isomorphic to $\mathbf{T}^n$. An n-dimensional torus $\mathbf{S}$ is called k-split if it is k-isomorphic to $\mathbf{T}^n$. If $\mathbf{G}$ is semisimple, then $\mathrm{rank}_k \mathbf{G}$ is defined as the dimension of a maximal k-split torus in $\mathbf{G}$. We say that $\mathbf{G}$ is *k-anisotropic* if $\mathrm{rank}_k \mathbf{G} = 0$.

For a semisimple linear algebraic group $\mathbf{G}$ defined over $\mathbb{Q}$, the *Borel–Harish-Chandra theorem* says that an arithmetic subgroup $\Gamma \leq \mathbf{G}$ is a lattice in the semisimple Lie group $\mathbf{G}(\mathbb{R})$ and the lattice is uniform if and only if $\mathbf{G}$ is $\mathbb{Q}$-anisotropic. Hence arithmetic groups provide a wealth of lattices in semisimple Lie groups. Margulis' seminal *arithmeticity theorem* asserts a partial converse: if a semisimple Lie group G maps with compact kernel and compact cokernel to $\mathbf{G}(\mathbb{R})$ for a connected semisimple linear algebraic $\mathbb{Q}$-group $\mathbf{G}$ with $\mathrm{rank}_{\mathbb{R}} \mathbf{G} \geq 2$, then the image of every *irreducible* lattice $\Gamma \leq G$ is conjugate to an arithmetic subgroup of $\mathbf{G}$. Here Γ is called *irreducible* if it is not virtually a product of lattices $\Gamma_1\Gamma_2$ coming from a nontrivial decomposition $G = G_1G_2$ with $G_1 \cap G_2$ central.

For extensive treatments of this material, the reader is referred to the monographs [146] by Platonov and Rapinchuk and [125] by Margulis. A more gentle introduction can be found in Witte Morris [175].

With all these new notions at hand, we can now formulate one of the most influential conjectures on torsion growth in homology in recent years [18, Conjecture 1.3]. Let $\mathbf{G}$ be a $\mathbb{Q}$-anisotropic semisimple linear algebraic group defined over $\mathbb{Q}$ and let $\Gamma \leq \mathbf{G}$ be a congruence subgroup. Consider an *algebraic* representation of $\mathbf{G}$ on a $\mathbb{Q}$-vector space V. Here "algebraic" means that a choice of a $\mathbb{Q}$-basis of V yields a $\mathbb{Q}$-morphism $\mathbf{G} \to \mathbf{GL_n}$. We fix a Γ-invariant $\mathbb{Z}$-lattice $M \subset V$, which always exists according to [146, Remark, p. 173]. Finally, let (Γ_i) be a decreasing chain of (not necessarily normal) congruence subgroups of Γ such that $\bigcap_{i \geq 0} \Gamma_i = \{1\}$.

The semisimple algebraic group $\mathbf{G}$ defines the semisimple Lie group $G = \mathbf{G}(\mathbb{R})$ for which the deficiency $\delta(G)$ and the symmetric space $X = G/K$ are defined as in Example 6.14. The Lie algebra $\mathfrak{k}$ of K defines the so-called *Cartan decomposition* $\mathfrak{g} = \mathfrak{k} \oplus \mathfrak{p}$ of the Lie algebra $\mathfrak{g}$ of G by defining the subspace $\mathfrak{p}$ as the orthogonal complement of $\mathfrak{k}$ with respect to the "Killing form" on $\mathfrak{g}$. This allows us to identify the tangent space $T_K X$ with $\mathfrak{p}$. Since the Killing form is positive definite on $\mathfrak{p}$, we obtain a G-invariant Riemannian metric on X by translation and hence a possible normalization of the volume $\mathrm{vol}(\Gamma \backslash X)$ of the "orbifold" $\Gamma \backslash X$ (which is a manifold if Γ is torsion-free).

Conjecture 6.27 (Bergeron–Venkatesh [18]) For every $n \geq 1$, there exists a constant $C_{n,G,M} \geq 0$ such that

$$\lim_{i \to \infty} \frac{\log |H_n(\Gamma_i; M)_{\mathrm{tors}}|}{[\Gamma : \Gamma_i]} = C_{n,G,M} \, \mathrm{vol}(\Gamma \backslash X)$$

and we have $C_{n,G,M} > 0$ if and only if $\delta(G) = 1$ and $\dim X = 2n + 1$.

Moreover, Bergeron and Venkatesh give an explicit description of the occurring positive constants $C_{n,G,M}$. Let us consider the case of a group $\mathbf{G}$ for which $\dim X$ is odd and suppose the chain (Γ_i) consists of normal and torsion-free congruence subgroups. If we choose $V = \mathbb{Q}$ to be the trivial one-dimensional representation, then of course $M = \mathbb{Z} \subset \mathbb{Q}$ is Γ-invariant and the Bergeron–Venkatesh conjecture makes the same prediction as Conjecture 6.22. Indeed, if $n = (\dim X - 1)/2$, then $(-1)^n C_{n,G,\mathbb{Z}}$ is Olbrich's constant mentioned in Example 6.14, so that the right hand side of the Bergeron–Venkatesh conjecture equals $|\rho^{(2)}(\Gamma)|$. In general, the right hand side accordingly has an interpretation as "twisted" ℓ^2-torsion.

Recall that in the previous section we extracted three key issues when trying to prove Conjecture 6.22. Two of these were the following. One needs to get rid of the regulators which relate torsion in homology with determinants; and to obtain convergence of these determinants, one would need to know that the cellular differentials on $\Gamma_i \backslash X$ do not have too many too small singular values. Both issues would go away if there was some $\varepsilon > 0$ such that for all $i \geq 0$ and all $n \geq 1$, the Laplacians on $C_n(\Gamma_i \backslash X; \mathbb{C})$ had spectrum within $[\varepsilon, \infty)$. For firstly, regulators $R_n(\Gamma_i \backslash X)$ do not occur if $H_n(\Gamma_i \backslash X; \mathbb{Z})_{\text{free}}$ is trivial and secondly, weak convergence of spectral measures now implies convergence of determinants because the last inequality in the proof of Proposition 5.47 becomes an equality as the logarithm is a bounded function on $[\varepsilon, 1)$. However, a well-known conjecture of Gromov, the *zero-in-the-spectrum conjecture* [63, Question 4.B.], asserts precisely that for a contractible cocompact Γ-manifold X, there should be at least one degree in which the spectrum of the ℓ^2-Laplacian contains zero. Consequently, the above condition would never be satisfied. Even worse, the zero-in-the-spectrum conjecture follows from the *strong Novikov conjecture* which is known to hold true for discrete subgroups of Lie groups [111, Corollary 4]. So in an arithmetic setting as we consider in this section, there is no hope to find manifolds with uniform spectral gap about zero in all degrees—as long as we are working with the trivial coefficient system $\mathbb{Z}$.

The point of this section is that there are however nontrivial coefficient systems M for which the above condition is satisfied and the Laplacians do have spectrum bounded away from zero. In fact, Bergeron–Venkatesh show that in the interesting case when $\delta(G) = 1$, such *strongly acyclic* Γ-modules M always exist [18, Section 8.1]. For these coefficient systems, they carry out the second and third step in the proof strategy for Conjecture 6.22 from Sect. 6.5 in the analytic setting: Strong acyclicity implies the vanishing of regulators so that the Cheeger–Müller theorem for unimodular representations [129] identifies integral torsion $\rho^M(\Gamma_i \backslash X)$, obtained from Definition 6.23 by using M instead of $\mathbb{Z}$, with the corresponding *analytic Ray–Singer torsion*. Strong acyclicity moreover rules out small eigenvalues of the differential form Laplacians in terms of which Ray–Singer torsion is defined. From this, Bergeron–Venkatesh conclude convergence of Ray–Singer torsion to *analytic ℓ^2-torsion with coefficients in M* in great generality: it is enough that the injectivity radius (see p. 107) of $\Gamma_i \backslash X$ tend to infinity. As pointed out in [2, Section 8.3], the proof is also easily adapted to the condition that $\Gamma_i \backslash X$ Benjamini–Schramm converges to X. Analytic ℓ^2-torsion is proportional to the volume of $\Gamma \backslash X$ because

G acts transitively by isometries on X. The sign of the proportionality constant is $(-1)^m$ for $\dim X = 2m + 1$ by an explicit computation as in Oblrich [139]. To sum up, we obtain

$$\lim_{i \to \infty} \sum_{n \geq 0} (-1)^n \frac{\log |H_n(\Gamma_i \backslash X; M)_{\text{tors}}|}{[\Gamma : \Gamma_i]} = (-1)^m C_{G,M} \operatorname{vol}(\Gamma \backslash X) \qquad (6.2)$$

with $C_{G,M} > 0$. To conclude the Bergeron–Venkatesh conjecture for strongly acyclic M, it would still remain to resolve the third issue, the "torsion Singer problem" that in fact only the middle degree summand produces exponential torsion growth. But at least we can drop the negative summands and get the following result.

Theorem 6.28 (Bergeron–Venkatesh [18]) *We have*

$$\liminf_{i \to \infty} \sum_{n \equiv m \, (2)} \frac{\log |H_n(\Gamma_i; M)_{\text{tors}}|}{[\Gamma : \Gamma_i]} \geq C_{G,M} \operatorname{vol}(\Gamma \backslash X).$$

In particular, exponential torsion growth occurs in some degree of the same parity as $(\dim X - 1)/2$. Moreover, one can see without too much trouble that both $H_0(\Gamma_i; M)_{\text{tors}}$ and $H_{\dim X - 1}(\Gamma_i; M)_{\text{tors}}$ grow at most polynomially [18, Section 8.6] in $[\Gamma : \Gamma_i]$. Therefore (6.2) implies more than Theorem 6.28 in low dimensional examples. If $\mathbf{G}$ satisfies $\mathbf{G}(\mathbb{R}) \cong_{\mathbb{R}} \operatorname{SL}(2; \mathbb{C})$, then $X = \operatorname{SL}(2; \mathbb{C})/\operatorname{SU}(2)$ is isometric to hyperbolic 3-space $\mathbb{H}^3$ and we get

$$\lim_{i \to \infty} \frac{\log |H_1(\Gamma_i; M)_{\text{tors}}|}{[\Gamma : \Gamma_i]} = C_{G,M} \operatorname{vol}(\Gamma \backslash X).$$

If $\mathbf{G}$ satisfies $\mathbf{G}(\mathbb{R}) \cong \operatorname{SL}(3; \mathbb{R})$, then $X = \operatorname{SL}(3; \mathbb{R})/\operatorname{SO}(3)$ is five-dimensional and we still get

$$\liminf_{i \to \infty} \frac{\log |H_2(\Gamma_i; M)_{\text{tors}}|}{[\Gamma : \Gamma_i]} \geq C_{G,M} \operatorname{vol}(\Gamma \backslash X).$$

In both cases the constant $C_{G,M}$ is positive so that we observe exponential torsion growth. As opposed to the case $\mathbf{G}(\mathbb{R}) \cong \operatorname{SL}(2; \mathbb{C})$, in the second case the condition $\mathbf{G}(\mathbb{R}) = \operatorname{SL}(3; \mathbb{R})$ implies that $\operatorname{rank}_{\mathbb{R}} \mathbf{G} = 2$, so that a well-known conjecture of Serre [146, (9.45), p. 556], says that $\mathbf{G}$ should have CSP. This would allow to control the growth of $H_1(\Gamma_i; M)_{\text{tors}}$ and $H_3(\Gamma_i; M)_{\text{tors}}$ as well, so that we would also get

$$\lim_{i \to \infty} \frac{\log |H_2(\Gamma_i; M)_{\text{tors}}|}{[\Gamma : \Gamma_i]} = C_{G,M} \operatorname{vol}(\Gamma \backslash X).$$

But unfortunately, $\mathbb{Q}$-anisotropic arithmetic lattices in $\operatorname{SL}(3; \mathbb{R})$ is one of the notorious open cases in Serre's conjecture. In contrast, the $\mathbb{Q}$-isotropic arithmetic lattice $\operatorname{SL}(3; \mathbb{Z}) \leq \operatorname{SL}(3; \mathbb{R})$ is well-known to have CSP. For Theorem 6.28, we needed a $\mathbb{Q}$-anisotropic group to obtain compact locally symmetric spaces $\Gamma_i \backslash X$ to which the Cheeger–Müller theorem implies. But one might anyway hope to obtain the conclusion of the Bergeron–Venkatesh conjecture for $\Gamma = SL(3; \mathbb{Z})$ with trivial coefficients $M = \mathbb{Z}$ and any sequence of distinct finite index subgroups (Γ_i) because CSP implies Benjamini–Schramm convergence of the quotients $\Gamma_i \backslash X$ to X in a strong sense [2, Section 5]. As stated in [16, Conjecture 5.1], we would then obtain the curious formula

$$\lim_{i \to \infty} \frac{\log |H_2(\Gamma_i; \mathbb{Z})_{\text{tors}}|}{[\Gamma : \Gamma_i]} = \frac{\zeta(3)}{96\sqrt{3}\pi^2}.$$

The value $\zeta(3) = 1.202\ldots$ of the Riemann zeta function is known as *Apéry's constant*, and enters as part of the volume computation for the locally symmetric space $SL(3; \mathbb{Z}) \backslash SL(3; \mathbb{R})/SO(3)$. Note moreover, that in the hyperbolic case, Müller–Pfaff extended the results of Bergeron–Venkatesh to non-uniform lattices [134].

As of now, it seems that all noteworthy positive results on exponential torsion growth hinge on the existence of strongly acyclic modules. On their construction, let us only say that the condition that the Γ-module M extends to an algebraic **G**-representation V has the virtue that the latter are well understood and classified (over $\mathbb{C}$) by so called "highest weights": the elements lying in a certain cone of the character lattice of **G**. Starting from a highest weight representation one can then construct strongly acyclic Γ-modules M if the highest weight lies outside a finite union of hyperplanes in the character space (which of course excludes the trivial representation). In this sense it is fair to say that **G** possesses a large supply of strongly acyclic representations.

Şengün [162, 163] has tested the Bergeron–Venkatesh conjecture numerically in the case of Bianchi groups $\Gamma = PSL(2; O_d)$ where d is a positive square-free integer and O_d is the ring of integers in the imaginary quadratic number field $\mathbb{Q}(\sqrt{-d})$. For prime ideals $\mathfrak{p} \subset O_d$ of residue degree one, he considers the arithmetic subgroup $\Gamma_0(\mathfrak{p})$ of those elements in Γ which reduce to an upper triangular (2×2)-matrix mod $\mathfrak{p}$. In the case of the trivial coefficient system $M = \mathbb{Z}$, and for $\mathfrak{p}$ of growing norm, the ratio

$$\frac{\log |H_1(\Gamma_0(\mathfrak{p}); \mathbb{Z})_{\text{tors}}|}{\text{vol}(\Gamma_0(\mathfrak{p}) \backslash \mathbb{H}^3)}$$

does indeed come close to the value $1/6\pi \approx 0.053\ldots$ as one would expect from the Bergeron–Venkatesh philosophy (not from the conjecture itself as the groups are again not cocompact). In non-arithmetic hyperbolic tetrahedral groups, however, Şengün considers similar subgroups $\Gamma_0(\mathfrak{p})$ for which the above ratio only comes close to $1/6\pi$ if $H_1(\Gamma_0(\mathfrak{p}); \mathbb{Z})$ is completely torsion. Otherwise, it is much smaller which suggests that the arithmetic setting in the Bergeron–Venkatesh conjecture is important to assure that the regulator contributions become small. Some more remarks in this direction can be found in [18, Conjecture 9.2 and below].

Let us finally mention that instead of investigating torsion when fixing a coefficient module $M \subset V$ and varying Γ, one can also ask to quantify the torsion growth if Γ is fixed and $M \subset V$ varies through rays of highest weight representations V. Once again, exponential torsion growth can be detected in this setup as the reader can learn from Marshall, Müller, and Pfaff in [126, 130–133]. Torsion in homology has recently received additional interest because of Scholze's work on the existence of Galois representations associated with mod p classes in the cohomology of locally symmetric spaces for **GL$_n$** over totally real or CM fields [160]. For a readable overview of the various ramifications of the material of this section, the reader is referred to the survey article [16].

6.7   Profiniteness Questions

The main value of Conjecture 6.22 is that it supplements the homology growth prediction in Conjecture 6.21 with a statement about torsion. But additionally, it has a neat and not quite obvious application to a question in group theory and 3-manifolds which we want to present in this section. Part of this material has previously appeared in the preprint [88].

With any group G we can associate the *profinite completion* defined as

$$\widehat{G} = \varprojlim_{N \trianglelefteq G,\ [G:N]<\infty} G/N,$$

the projective limit over the inverse system of all finite quotients of G. Hence $\widehat{G}$ is a compact, totally disconnected group (a *profinite* group). Totally disconnected spaces are T_1 and topological groups are T_2 if they are T_1, so profinite groups, in particular profinite completions of groups, are Hausdorff. The profinite completion comes with a canonical homomorphism $G \to \widehat{G}$ with dense image. Since all the projections $G \to G/N$ factor through $G \to \widehat{G}$, we see that $G \to \widehat{G}$ is injective if and only if G is residually finite. As every profinite group P is the projective limit of the quotients P/N by open (hence finite index) normal subgroups $N \trianglelefteq P$, we observe the universal property that every morphism $G \to P$ to a profinite group P factorizes uniquely through $G \to \widehat{G}$. We conclude that passing to the profinite completion is a functor: Given a group homomorphism $f : G \to H$, we compose it with $H \to \widehat{H}$ so that the universal property induces a continuous homomorphism $\widehat{f} : \widehat{G} \to \widehat{H}$ covering f.

The easiest examples of profinite completions are $\widehat{F} = F$ if F is finite and $\widehat{\mathbb{Z}} = \prod_p \mathbb{Z}_p$ by the Chinese remainder theorem, where $\mathbb{Z}_p$ denotes the p-adic integers. Moreover:

Lemma 6.29 *We have a canonical isomorphism* $\Phi \colon \widehat{G \times H} \xrightarrow{\cong} \widehat{G} \times \widehat{H}$.

Proof Consider the three canonical homomorphisms $\phi_G \colon G \to \widehat{G}$, $\phi_H \colon H \to \widehat{H}$, and $\kappa \colon G \times H \to \widehat{G \times H}$. The first two yield the product morphism $\phi \colon G \times H \to \widehat{G} \times \widehat{H}$ and hence the universal property provides a morphism $\Phi \colon \widehat{G \times H} \to \widehat{G} \times \widehat{H}$ such that $\phi = \Phi \circ \kappa$. Conversely, the canonical inclusions $\iota_G \colon G \to G \times H$ and $\iota_H \colon H \to G \times H$ are split injective, so we obtain embeddings $\widehat{\iota_G} \colon \widehat{G} \to \widehat{G \times H}$ and $\widehat{\iota_H} \colon \widehat{H} \to \widehat{G \times H}$ satisfying $\kappa \circ \iota_G = \widehat{\iota_G} \circ \phi_G$ and $\kappa \circ \iota_H = \widehat{\iota_H} \circ \phi_H$. Therefore, sending $(x, y) \in \widehat{G} \times \widehat{H}$ to the commutator of $\widehat{\iota_G}(x)$ and $\widehat{\iota_H}(y)$ defines a continuous map $\widehat{G} \times \widehat{H} \to \widehat{G \times H}$ that is constantly 1 on the dense subset $\phi(G \times H)$. Hence it is constantly 1 everywhere and $\widehat{\iota_G}(\widehat{G})$ and $\widehat{\iota_H}(\widehat{H})$ commute. This shows that sending (x, y) to $\widehat{\iota_G}(x) \cdot \widehat{\iota_H}(y)$ defines a homomorphism $\Psi \colon \widehat{G} \times \widehat{H} \to \widehat{G \times H}$ satisfying $\kappa = \Psi \circ \phi$. Thus $\Psi \circ \Phi$ restricts to the identity on the dense subgroup $\kappa(G \times H)$ and $\Phi \circ \Psi$ restricts to the identity on the dense subgroup $\phi(G \times H)$. By continuity, Φ and Ψ are inverses of one another. $\qquad\square$

Now suppose two groups G and K are *profinitely isomorphic*, meaning $\widehat{G} \cong \widehat{K}$ as topological groups. Does it follow that $G \cong K$? The answer is "never", because Higman's group H from p. 88 has $\widehat{H} = \{1\}$, hence $\widehat{G \times H} \cong \widehat{G}$ for all G by Lemma 6.29. However, if we assume that G and K are residually finite, the question becomes interesting.

Definition 6.30 A finitely generated, residually finite group G is *profinitely rigid* if for every finitely generated, residually finite group K with $\widehat{K} \cong \widehat{G}$, we have $K \cong G$.

In this definition, it makes no difference whether "$\widehat{K} \cong \widehat{G}$" means topological or abstract isomorphism because assuming the groups are finitely generated has the effect that any abstract isomorphism $\widehat{K} \cong \widehat{G}$ is also a homeomorphism. This is an immediate consequence of a deep theorem due to Nikolov and Segal [137]. To get acquainted with the definition, let us convince ourselves that finitely generated abelian groups are profinitely rigid. To this end, we show the stronger statement that the abelianization is a *profinite invariant* or, for short, is *profinite*: profinitely isomorphic groups have isomorphic abelianizations [151, Proposition 3.2]. From now on, G and K shall denote finitely generated and residually finite groups.

Proposition 6.31 *If the group K embeds densely into $\widehat{G}$, then there exists an epimorphism $H_1(K) \to H_1(G)$.*

Proof Let p be a prime number which does not divide the group order $|H_1(K)_{\mathrm{tors}}|$ and let us set $r = \dim_{\mathbb{Q}} H_1(G; \mathbb{Q})$. It is apparent that we have an epimorphism $G \to (\mathbb{Z}/p\mathbb{Z})^r \oplus H_1(G)_{\mathrm{tors}}$. By the universal property, this epimorphism extends uniquely to an epimorphism $\widehat{G} \to (\mathbb{Z}/p\mathbb{Z})^r \oplus H_1(G)_{\mathrm{tors}}$. Since K embeds densely into $\widehat{G}$, the latter map restricts to an epimorphism $K \to (\mathbb{Z}/p\mathbb{Z})^r \oplus H_1(G)_{\mathrm{tors}}$. This epimorphism must lift to an epimorphism $K \to \mathbb{Z}^r \oplus H_1(G)_{\mathrm{tors}} \cong H_1(G)$ because p is coprime to $|H_1(K)_{\mathrm{tors}}|$. The latter epimorphism factors through $H_1(K)$. $\qquad\square$

Corollary 6.32 *If we have $\widehat{G} \cong \widehat{K}$, then $H_1(G) \cong H_1(K)$.*

Proof Since there exist surjections in both directions the groups $H_1(G)$ and $H_1(K)$ have the same free abelian rank. Thus either surjection restricts to an isomorphism of the free parts and thus induces a surjection of the finite torsion quotients—which then must be a bijection.   □

While this shows that finitely generated abelian groups are profinitely rigid, already some virtually cyclic ones are not [151, Theorem 3.3]. Recently, it was shown with some effort that the figure eight knot group is profinitely rigid among all 3-manifold groups [25]. In general, however, profinite rigidity of fundamental groups of hyperbolic 3-manifolds even among themselves is open and appears out of reach for now. An at least formally easier but still open problem is the following.

Conjecture 6.33 Let M and N be closed, connected, orientable, irreducible 3-manifolds with infinite fundamental groups. Then $\widehat{\pi_1 M} \cong \widehat{\pi_1 N}$ implies vol $M =$ vol N.

The definition of irreducibility was given in Example 6.13 where we also reported that M and N have a unique geometric decomposition. Volume is defined as the sum of the volumes of the hyperbolic pieces in this decomposition. This is an invariant of the fundamental group only because M and N are aspherical by the sphere theorem [69, Theorem 4.3, p. 40] and Thurston geometrization also proves the *Borel conjecture* in dimension 3 as is surveyed in [98, Theorem 0.7]. So the conjecture claims that volume is profinite among 3-manifold groups. Conjecture 6.33 is again not intrinsically concerned with ℓ^2-invariants. But ℓ^2-methods might prove it.

Theorem 6.34 *Conjecture 6.22 implies Conjecture 6.33.*

The contrapositive of Theorem 6.34 says that constructing two profinitely isomorphic 3-manifold groups with differing covolume would disprove Conjecture 6.22. Funar [52] and Hempel [71] constructed examples of closed 3-manifolds with non-isomorphic but profinitely isomorphic fundamental groups. These examples carry *Sol* and $\mathbb{H}^2 \times \mathbb{R}$ geometry, respectively, and thus all have zero volume by definition. Wilkes [172] showed that Hempel's examples are the only ones among Seifert-fiber spaces. No examples with $\mathbb{H}^3$-geometry are known and as mentioned above, a first step in the direction that there should be no such examples was undertaken by Bridson and Reid [25] who showed that the figure eight knot group is profinitely rigid among 3-manifold groups.

We prepare the proof of Theorem 6.34 with a couple of propositions loosely following A. Reid's survey [151] while filling in some more details. Recall that the subspace topology of G in $\widehat{G}$ is called the *profinite topology*.

Proposition 6.35 *A subgroup $H \leq G$ is open in the profinite topology if and only if H has finite index in G.*

Proof By definition, $\widehat{G}$ carries the coarsest topology under which the projections $\widehat{G} \to G/G_i$ for finite index normal subgroups $G_i \trianglelefteq G$ are continuous. Since the compositions $G \to \widehat{G} \to G/G_i$ are the canonical projections, it follows that a subbase for the subspace topology of $G \subset \widehat{G}$ is given by the cosets of finite index normal subgroups of G.

If H has finite index in G, then so does the normal core $N = \bigcap_{g \in G} g^{-1} H g$. Thus $H = \bigcup_{h \in H} hN$ is open. Conversely, let $H \leq G$ be open. Then H is a union of finite intersections of finite index normal subgroups of G. In particular, H contains a finite index subgroup, so H has finite index itself. $\qquad\square$

Recall that open subgroups of $\widehat{G}$ have finite index because the cosets disjointly cover the compact space $\widehat{G}$. Though we will not use it, we remark that the Nikolov–Segal theorem cited above concludes from the finite generation of G that also the converse is true: finite index subgroups in $\widehat{G}$ are open.

Proposition 6.36 *Taking closure $H \mapsto \overline{H}$ in $\widehat{G}$ defines a 1-1-correspondence from the open subgroups of G to the open subgroups of $\widehat{G}$. The inverse is given by intersection $H \mapsto H \cap G$ with G. The correspondence $H \mapsto \overline{H}$ preserves the index, sends a normal subgroup $N \trianglelefteq G$ to a normal subgroup $\overline{N} \trianglelefteq \widehat{G}$, and in the latter case we have $\widehat{G}/\overline{N} \cong G/N$.*

The proof is given in [152, Prop. 3.2.2, p. 84]. Here is an easy consequence.

Corollary 6.37 *For $H_1, H_2 \leq G$ of finite index we have $\overline{H_1 \cap H_2} = \overline{H_1} \cap \overline{H_2}$.*

Proof By Propositions 6.35 and 6.36, the subgroups $\overline{H_1}$ and $\overline{H_2}$ are open in $\widehat{G}$ and

$$(\overline{H_1} \cap \overline{H_2}) \cap G = (\overline{H_1} \cap G) \cap (\overline{H_2} \cap G) = H_1 \cap H_2.$$

Applying Proposition 6.36 again yields $\overline{H_1} \cap \overline{H_2} = \overline{H_1 \cap H_2}$. $\qquad\square$

Recall from the proof of Theorem 5.29 that for a finitely generated, residually finite group G, we obtain a canonical choice of a chain

$$G = K_1 \geq K_2 \geq K_3 \geq \cdots$$

of finite index normal subgroups $K_i \trianglelefteq G$ satisfying $\bigcap_{i=1}^{\infty} K_i = \{1\}$ by defining K_i as the intersection of the normal subgroups of G with index at most i.

Proposition 6.38 *The intersection $\bigcap_{i=1}^{\infty} \overline{K_i}$ is trivial.*

Proof By the last two results, $\overline{K_i}$ is the intersection of all open normal subgroups of $\widehat{G}$ of index at most i. Thus $\bigcap_{i=1}^{\infty} \overline{K_i}$ is the intersection of all open subgroups and the proposition just says $\widehat{G}$ is T_1, or equivalently $\{e\}$ is closed in $\widehat{G}$. But this is true because $\{e\}$ is even a connected component. $\qquad\square$

Before we give the proof of Theorem 6.34, we put down one more observation.

Proposition 6.39 *If $H \leq G$ has finite index, then the identity on H extends uniquely to an isomorphism $\eta \colon \widehat{H} \to \overline{H}$ of topological groups.*

Proof A unit neighborhood basis of H in the subspace topology of $\overline{H}$ is given by the finite index normal subgroups $K_i \cap H$ where K_i are the finite index normal subgroups of G from above. By Proposition 6.35, a unit neighborhood basis of the profinite topology of H is given by the finite index normal subgroups of H. But since H has finite index in G, every finite index normal subgroup $K \trianglelefteq H$ still has finite index in G and therefore $K_i \leq K$ for i large enough. This shows that the two unit neighborhood bases define the same topology on H. So H embeds continuously and densely into $\widehat{H}$ and $\overline{H}$. As both $\widehat{H}$ and $\overline{H}$ are complete with respect to the canonical uniform structures, the identity on H extends uniquely to an isomorphism $\eta \colon \widehat{H} \to \overline{H}$ by the universal property of the uniform completion of topological groups. $\square$

Proof (of Theorem 6.34.) The groups $G = \pi_1 M$ and $H = \pi_1 N$ are finitely generated and residually finite, as a consequence of geometrization [70]. We fix an isomorphism $\widehat{G} \cong \widehat{H}$. Again, let $K_i \leq G$ be the intersection of all normal subgroups of G of index at most i. By Propositions 6.35 and 6.36, the group $L_i = H \cap \overline{K_i}$ is the intersection of all normal subgroups of H of index at most i and $[G : K_i] = [H : L_i]$. By Proposition 6.38 we have $\bigcap_i \overline{K_i} = \{1\}$ so that $\bigcap_i L_i = \{1\}$. From Proposition 6.39 we get $\widehat{K_i} \cong \widehat{L_i}$ so that Corollary 6.32 implies $|H_1(K_i)_{\text{tors}}| = |H_1(L_i)_{\text{tors}}|$. By Example 6.13 we have $\rho^{(2)}(\widetilde{M}) = -\operatorname{vol}(M)/6\pi$ for the aspherical manifold M and similarly for N. The manifolds $M_i = K_i \backslash \widetilde{M}$ and $N_i = L_i \backslash \widetilde{N}$ are aspherical, too, hence models for BK_i and BL_i. Therefore Conjecture 6.22, if true, implies

$$\operatorname{vol}(M) = 6\pi \lim_{i \to \infty} \frac{\log |H_1(K_i)_{\text{tors}}|}{[G : K_i]} = 6\pi \lim_{i \to \infty} \frac{\log |H_1(L_i)_{\text{tors}}|}{[H : L_i]} = \operatorname{vol}(N). \quad \square$$

Theorem 6.34 also says that a proof of Conjecture 6.22 would make substantial progress on the profinite rigidity problem for hyperbolic 3-manifolds.

Corollary 6.40 *If Conjecture 6.22 is true, then profinite isomorphism classes of fundamental groups of closed hyperbolic 3-manifolds are finite.*

Proof For $n \geq 3$, there are only finitely many hyperbolic n-manifolds of any given volume. The case $n = 3$ is due to Jørgensen–Gromov–Thurston [165, Corollary 6.6.2]. (The case $n \geq 4$ and in fact the corresponding statement for all closed locally symmetric spaces of noncompact type without $\mathbb{H}^2$ or $\mathbb{H}^3$ factors is due to Wang [170, Theorem 8.1].) $\square$

Generalizing Conjecture 6.33, one might dare and ask whether actually ℓ^2-torsion is profinite among ℓ^2-acyclic, residually finite groups with finite classifying space. It turns out that this is overly optimistic [91].

Theorem 6.41 *There exist profinitely isomorphic ℓ^2-acyclic, residually finite groups G_1, G_2, and G_3 which have finite models for EG_i and satisfy*

$$\rho^{(2)}(G_1) < 0, \qquad \rho^{(2)}(G_2) = 0, \qquad \rho^{(2)}(G_3) > 0.$$

So not even the sign of the ℓ^2-torsion is a profinite invariant. The theorem is actually an easy corollary of the corresponding statement for the even-dimensional cousin of ℓ^2-torsion [91, Theorem 1.3].

Theorem 6.42 *There exist profinitely isomorphic, residually finite groups G_1, G_2, and G_3 which have finite models for EG_i and satisfy*

$$\chi(G_1) < 0, \qquad \chi(G_2) = 0, \qquad \chi(G_3) > 0.$$

The proof proceeds roughly as follows. The special orthogonal group $\mathbf{SO}(q)$ of an integral quadratic form q has a simply connected covering in the sense of algebraic groups denoted by $\mathbf{Spin}(q)$, the *spinor group* of q. Of special interest are the groups $\mathbf{Spin}(r, s)$ where q is diagonal and has r times the value "+1" and s times the value "−1" on the diagonal. Standard quadratic form theory reveals that the quaternary forms

$$x_1^2 + x_2^2 + x_3^2 + x_4^2 \quad \text{and} \quad -x_1^2 - x_2^2 - x_3^2 - x_4^2$$

are isometric over $\mathbb{Z}_p$ for each (finite) prime p. This and two deep results in arithmetic groups, namely *strong approximation* [146, Chapter 7] and the congruence subgroup property introduced on p. 151, have the effect that the arithmetic groups given by the $\mathbb{Z}$-points $\mathbf{Spin}(8, 2)(\mathbb{Z})$ and $\mathbf{Spin}(4, 6)(\mathbb{Z})$ are profinitely isomorphic. Yet they are not isomorphic themselves as a consequence of *strong rigidity* [128]. M. Aka introduced this trick and applied it similarly to come up with two profinitely isomorphic groups with and without *Kazhdan's property* (T) [5]. The torsion-free congruence subgroups $G_{8,2}$ and $G_{4,6}$ given by the kernels of

$$\mathbf{Spin}(8, 2)(\mathbb{Z}) \to \mathbf{Spin}(8, 2)(\mathbb{Z}/4) \quad \text{and} \quad \mathbf{Spin}(4, 6)(\mathbb{Z}) \to \mathbf{Spin}(4, 6)(\mathbb{Z}/4)$$

are still profinitely but not honestly isomorphic. Working with $G_{8,2}$ and $G_{4,6}$ also avoids some technicalities when applying Kionke's adelic version of Harder's Gauss–Bonnet formula [95, Theorem 3.3] to compute the Euler characteristics of $G_{8,2}$ and $G_{4,6}$. It turns out that

$$\chi(G_{8,2}) = 2^{89} \, 5^2 \, 17 \quad \text{and} \quad \chi(G_{4,6}) = 2^{90} \, 5^2 \, 17$$

so that the two Euler characteristics differ by a factor of two. This shows that the absolute value of the Euler characteristic is not a profinite invariant. But neither is the sign because setting $c = 2^{89} \, 5^2 \, 17$, we can define G_1, G_2, and G_3 as the free product of the free group F_{2c^2} with

$$G_{8,2} \times G_{8,2}, \qquad G_{8,2} \times G_{4,6}, \quad \text{and} \quad G_{4,6} \times G_{4,6},$$

respectively. The groups G_i are still pairwise profinitely isomorphic since the profinite completion functor preserves products and coproducts. The Euler characteristic is multiplicative for products and additive for pushouts, so

$$\chi(G_1) = -c^2, \quad \chi(G_2) = 0, \quad \text{and} \quad \chi(G_3) = 2c^2,$$

showing Theorem 6.42. To deduce Theorem 6.41, take a closed hyperbolic 3-manifold M, replace G_i by $\pi_1 M \times G_{4-i}$, and apply Theorem 6.9 (iv) and (vi).

Because of the Euler–Poincaré formula 3.19, Theorem 6.42 implies that ℓ^2-Betti numbers cannot generally be profinite either. Notwithstanding:

Theorem 6.43 *The first ℓ^2-Betti number is profinite among finitely presented, residually finite groups.*

This was observed by Bridson et al. [26, Corollary 3.3] to be a consequence of Lück's approximation theorem and gave the blueprint for the proof of Theorem 6.34. It was moreover already mentioned in [151, 6.2, p. 88] that the spinor groups considered by Aka in [5], show that some higher ℓ^2-Betti numbers are not profinite. Using *S-arithmetic groups* one can improve the construction to see that actually no higher ℓ^2-Betti number is profinite. This follows from the following result [90, Theorem 1].

Theorem 6.44 *For $k \geq 2$, let $p_1, \ldots, p_k$ be different primes from the arithmetic progression $89 + 24\mathbb{N}$. Consider the two S-arithmetic groups*

$$G_\pm^k = \mathbf{Spin}(\langle \pm 1, \pm 1, \pm 1, \pm p_1 \cdots p_k, 3 \rangle) \left(\mathbb{Z}[\tfrac{1}{p_1 \cdots p_k}] \right).$$

Then the groups G_+^k and G_-^k are profinitely isomorphic and

- $b_n^{(2)}(G_+^k) > 0$ *if and only if $n = k$,*
- $b_n^{(2)}(G_-^k) > 0$ *if and only if $n = k + 2$.*

Let us point out that the theorem is meaningful because 89 and 24 are coprime so that Dirichlet's theorem ensures the progression contains infinitely many primes. The letter "*S*" in *S-arithmetic* refers to the set $S = \{p_1, \ldots, p_k\}$ of prime numbers we allow to invert. The *S*-arithmetic groups appearing in the theorem are finitely presented [146, Theorem 5.11, p. 272] and residually finite because they are linear. So not even the property $b_n^{(2)}(G) = 0$ is profinite for $n \geq 2$ among finitely presented, residually finite groups. Note however that the groups $G_\pm^k$ from the theorem have nonzero ℓ^2-Betti numbers in degrees which differ by two. This leaves open the option that the sign of the Euler characteristic is profinite among *S*-arithmetic groups, even though we have seen that the absolute value of the Euler characteristic is not, and neither is the sign among the more general class of residually finite groups with finite classifying space. For arithmetic groups we have indeed a positive result [91].

Theorem 6.45 *Let $\mathbf{G_1}$ and $\mathbf{G_2}$ be linear algebraic groups over number fields k_1 and k_2 and let $\Gamma_1 \leq \mathbf{G_1}$ and $\Gamma_2 \leq \mathbf{G_2}$ be arithmetic subgroups. Assume that $\mathbf{G_1}$ and*

$\mathbf{G_2}$ *have CSP and* Γ_1 *and* Γ_2 *are profinitely commensurable. Then* $\text{sign}\,\chi(\Gamma_1) = \text{sign}\,\chi(\Gamma_2)$.

Here *profinitely commensurable* means the groups have finite index subgroups which are profinitely isomorphic. As usual, $\text{sign}(x)$ takes the values $-1, 0, 1$, for $x < 0, x = 0, x > 0$, respectively. We remind the reader of Serre's conjecture [146, (9.45), p. 556], already mentioned on p. 154, which says CSP is granted if the group has "higher rank".

The Singer conjecture is known for arithmetic groups $\Gamma \leq \mathbf{G}$ in the sense that they have a non-zero ℓ^2-Betti number in at most one degree which would be the middle dimension of the associated symmetric space if $\mathbf{G}$ is semisimple. Thus, Theorem 6.45 shows that being ℓ^2-acyclic is profinite for arithmetic groups with CSP. In particular, it makes sense to ask for profiniteness of the sign of χ's cousin $\rho^{(2)}$ and the following companion to Theorem 6.45 is obtained in [91, Theorem 1.4].

Theorem 6.46 *In addition to the assumptions of Theorem 6.45, assume that* $\text{rank}_{k_1}(\mathbf{G}_1) = \text{rank}_{k_2}(\mathbf{G}_2) = 0$ *and that* Γ_1 *(equivalently* Γ_2*) is* ℓ^2*-acyclic. Then* $\text{sign}\,\rho^{(2)}(\Gamma_1) = \text{sign}\,\rho^{(2)}(\Gamma_2)$.

Generalizing the situation on p. 152 where $k_i = \mathbb{Q}$, the assumption of $\mathbf{G}_i$ being k_i-anisotropic effects that the arithmetic groups Γ_i are uniform lattices in the Lie groups $G_i = \mathbf{G}_i(\mathbb{R})^{r_i} \times \mathbf{G}_i(\mathbb{C})^{s_i}$ where r_i and s_i is the number of real embeddings and pairs of conjugate complex embeddings of k_i, respectively. We conjecture that this assumption is not needed so that the result also holds for non-uniform Γ_i. But in view of the discussion in Example 6.15, dropping the uniformity assumption would either require proving the Lück–Sauer–Wegner proportionality conjecture for ℓ^2-torsion or finding any other way to compare cellular and analytic ℓ^2-torsion of non-uniform lattices. At least it follows again from [84, Theorem 1.2] that the non-uniform extension of the theorem holds true if one of the Lie groups G_i has even deficiency.

Profinite rigidity remains a rapidly developing field with numerous challenging open problems. For further reading, we recommend the introductory overview [151] by Reid as well as Nikolov's informative survey [136] on general algebraic properties of profinite groups.

References

1. M. Abért, N. Nikolov, Rank gradient, cost of groups and the rank versus Heegaard genus problem. J. Eur. Math. Soc. **14**(5), 1657–1677 (2012). MR 2966663
2. M. Abért, N. Bergeron, I. Biringer, T. Gelander, N. Nikolov, J. Raimbault, I. Samet, On the growth of L^2-invariants for sequences of lattices in Lie groups. Ann. Math. (2) **185**(3), 711–790 (2017). MR 3664810
3. I. Agol, Criteria for virtual fibering. J. Topol. **1**(2), 269–284 (2008). MR 2399130
4. I. Agol, The virtual Haken conjecture. Doc. Math. **18**, 1045–1087 (2013). With an appendix by Agol, Daniel Groves, and Jason Manning. MR 3104553
5. M. Aka, Profinite completions and Kazhdan's property (T). Groups Geom. Dyn. **6**(2), 221–229 (2012). MR 2914858
6. R.C. Alperin, An elementary account of Selberg's lemma. Enseign. Math. (2) **33**(3–4), 269–273 (1987). MR 925989
7. M. Aschenbrenner, S. Friedl, H. Wilton, *3-Manifold Groups*. EMS Series of Lectures in Mathematics (European Mathematical Society, Zürich, 2015). MR 3444187
8. M. Atiyah, Elliptic operators, discrete groups and von Neumann algebras, in *Colloque "Analyse et Topologie" en l'Honneur de Henri Cartan (Orsay, 1974)* (Society of Mathematics, Paris, 1976), pp. 43–72. Astérisque, No. 32-33. MR 0420729
9. M. Atiyah, F. Hirzebruch, Spin-manifolds and group actions, in *Essays on Topology and Related Topics (Mémoires dédiés à Georges de Rham)* (Springer, New York, 1970), pp. 18–28. MR 0278334
10. T. Austin, Rational group ring elements with kernels having irrational dimension. Proc. Lond. Math. Soc. (3) **107**(6), 1424–1448 (2013). MR 3149852
11. H. Bass, J. Milnor, J.-P Serre, Solution of the congruence subgroup problem for $SL_n (n \geq 3)$ and $Sp_{2n} (n \geq 2)$. Inst. Hautes Études Sci. Publ. Math. **33**, 59–137 (1967). MR 0244257
12. H. Bauer, *Maß- und Integrationstheorie*, 2nd edn. (de Gruyter Lehrbuch. [de Gruyter Textbook], Walter de Gruyter, Berlin, 1992) (German). MR 1181881
13. B. Bekka, A. Valette, Group cohomology, harmonic functions and the first L^2-Betti number. Potential Anal. **6**(4), 313–326 (1997). MR 1452785
14. B. Bekka, P. de la Harpe, A. Valette, *Kazhdan's Property (T)*. New Mathematical Monographs, vol. 11 (Cambridge University Press, Cambridge, 2008). MR 2415834
15. F. Ben Aribi, The L^2-Alexander invariant detects the unknot. Ann. Sc. Norm. Super. Pisa Cl. Sci. (5) **15**, 683–708 (2016). MR 3495444
16. N. Bergeron, Torsion homology growth in arithmetic groups, in *European Congress of Mathematics* (Eur. Math. Soc., Zürich, 2018), pp. 263–287. MR 3887771

© Springer Nature Switzerland AG 2019

H. Kammeyer, *Introduction to ℓ²-invariants*, Lecture Notes in Mathematics 2247,
https://doi.org/10.1007/978-3-030-28297-4

17. N. Bergeron, D. Gaboriau, Asymptotique des nombres de Betti, invariants l^2 et laminations. Comment. Math. Helv. **79**(2), 362–395 (2004) (French, with English summary). MR 2059438

18. N. Bergeron, A. Venkatesh, The asymptotic growth of torsion homology for arithmetic groups. J. Inst. Math. Jussieu **12**(2), 391–447 (2013). MR 3028790

19. N. Bergeron, Torsion homology growth in arithmetic groups, in *European Congress of Mathematics* (Eur. Math. Soc., Zürich, 2018), pp. 263–287. MR 3887771

20. B. Blackadar, *Operator Algebras*. Encyclopaedia of Mathematical Sciences, vol. 122 (Springer, Berlin, 2006). Theory of C^*-algebras and von Neumann algebras; Operator Algebras and Non-commutative Geometry, III. MR 2188261

21. A. Borel, The L^2-cohomology of negatively curved Riemannian symmetric spaces. Ann. Acad. Sci. Fenn. Ser. A I Math. **10**, 95–105 (1985). MR 802471

22. A. Borel, J.-P. Serre, Corners and arithmetic groups. Comment. Math. Helv. **48**, 436–491 (1973). Avec un appendice: Arrondissement des variétés à coins, par A. Douady et L. Hérault. MR 0387495

23. M. Borinsky, K. Vogtmann, The Euler characteristic of $\mathrm{Out}(F_n)$. https://arxiv.org/abs/1907.03543

24. N. Bourbaki, *Elements of Mathematics. Algebra, Part I: Chapters 1–3* (Addison-Wesley, Reading, 1974). Translated from the French. MR 0354207

25. M.R. Bridson, A.W. Reid, Profinite rigidity, fibering, and the figure-eight knot (2015). arXiv:1505.07886

26. M.R. Bridson, M.D.E. Conder, A.W. Reid, Determining Fuchsian groups by their finite quotients. Israel J. Math. **214**(1), 1–41 (2016). MR 3540604

27. S.D. Brodskiĭ, Equations over groups, and groups with one defining relation. Sibirsk. Mat. Zh. **25**(2), 84–103 (1984) (Russian). MR 741011

28. E.J. Brody, The topological classification of the lens spaces. Ann. Math. (2) **71**, 163–184 (1960). MR 0116336

29. J. Cheeger, M. Gromov, L_2-cohomology and group cohomology. Topology **25**(2), 189–215 (1986). MR 837621

30. M. Clay, ℓ^2-torsion of free-by-cyclic groups. Q. J. Math. **68**(2), 617–634 (2017). MR 3667215

31. D. Crowley, W. Lück, T. Macko, *Surgery Theory: Foundations* (to appear). http://www.mat.savba.sk/~macko/

32. W. Dicks, P.A. Linnell, L^2-Betti numbers of one-relator groups. Math. Ann. **337**(4), 855–874 (2007). MR 2285740

33. J. Dixmier, *von Neumann Algebras*. North-Holland Mathematical Library, vol. 27 (North-Holland, Amsterdam, 1981). With a preface by E.C. Lance; Translated from the second French edition by F. Jellett. MR 641217

34. J. Dodziuk, de Rham-Hodge theory for L^2-cohomology of infinite coverings. Topology **16**(2), 157–165 (1977). MR 0445560

35. J. Dodziuk, L^2 Harmonic forms on rotationally symmetric Riemannian manifolds. Proc. Am. Math. Soc. **77**(3), 395–400 (1979). MR 545603

36. J. Dodziuk, V. Mathai, Approximating L^2 invariants of amenable covering spaces: a combinatorial approach. J. Funct. Anal. **154**(2), 359–378 (1998). MR 1612713

37. J. Dodziuk, P. Linnell, V. Mathai, T. Schick, S. Yates, Approximating L^2-invariants and the Atiyah conjecture. Commun. Pure Appl. Math. **56**(7), 839–873 (2003). Dedicated to the memory of Jürgen K. Moser. MR 1990479

38. J. Dubois, C. Wegner, Weighted L^2-invariants and applications to knot theory. Commun. Contemp. Math. **17**(1), 1450010 (2015). MR 3291974

39. J. Dubois, S. Friedl, W. Lück, The L^2-Alexander torsion is symmetric. Algebr. Geom. Topol. **15**(6), 3599–3612 (2015). MR 3450772

40. J. Dubois, S. Friedl, W. Lück, The L^2-Alexander torsion of 3-manifolds. J. Topol. **9**(3), 889–926 (2016). MR 3551842

41. G. Elek, The strong approximation conjecture holds for amenable groups. J. Funct. Anal. **239**(1), 345–355 (2006). MR 2258227

42. G. Elek, E. Szabó, Hyperlinearity, essentially free actions and L^2-invariants. The sofic property. Math. Ann. **332**(2), 421–441 (2005). MR 2178069
43. J. Elstrodt, *Maß- und Integrationstheorie*, 4th edn. Springer Textbook (Springer-Lehrbuch; Springer, Berlin, 2005) (German). Grundwissen Mathematik. [Basic Knowledge in Mathematics]. MR 2257838
44. M. Farber, Geometry of growth: approximation theorems for L^2 invariants. Math. Ann. **311**(2), 335–375 (1998). MR 1625742
45. D.R. Farkas, P.A. Linnell, Congruence subgroups and the Atiyah conjecture, in *Groups, Rings and Algebras*. Contemporary Mathematics, vol. 420 (American Mathematical Society, Providence, 2006), pp. 89–102. MR 2279234
46. R.H. Fox, A quick trip through knot theory, in *Topology of 3-Manifolds and Related Topics: Proceedings of The University of Georgia Institute, 1961* (Prentice-Hall, Englewood Cliffs, 1962), pp. 120–167. MR 0140099
47. S. Friedl, T. Kim, The Thurston norm, fibered manifolds and twisted Alexander polynomials. Topology **45**(6), 929–953 (2006). MR 2263219
48. S. Friedl, T. Kitayama, The virtual fibering theorem for 3-manifolds. Enseign. Math. **60**(1–2), 79–107 (2014). MR 3262436
49. S. Friedl, W. Lück, The L^2-torsion function and the Thurston norm of 3-manifolds (2015). arXiv:1510.00264
50. S. Friedl, W. Lück, Universal L^2-torsion, polytopes and applications to 3-manifolds. Proc. Lond. Math. Soc. (3) **114**(6), 1114–1151 (2017). MR 3661347
51. S. Friedl, A. Juhász, J. Rasmussen, The decategorification of sutured Floer homology. J. Topol. **4**(2), 431–478 (2011). MR 2805998
52. L. Funar, Torus bundles not distinguished by TQFT invariants. Geom. Topol. **17**(4), 2289–2344 (2013). With an appendix by Funar and Andrei Rapinchuk. MR 3109869
53. A. Furman, A survey of measured group theory, in *Geometry, Rigidity, and Group Actions*. Chicago Lectures in Mathematics (University Chicago Press, Chicago, 2011), pp. 296–374. MR 2807836
54. D. Gaboriau, Coût des relations d'équivalence et des groupes. Invent. Math. **139**(1), 41–98 (2000) (French, with English summary). MR 1728876
55. D. Gaboriau, Invariants l^2 de relations d'équivalence et de groupes. Publ. Math. Inst. Hautes Études Sci. **95**, 93–150 (2002) (French). MR 1953191
56. D. Gaboriau, Orbit equivalence and measured group theory, in *Proceedings of the International Congress of Mathematicians*, vol. III (Hindustan Book Agency, New Delhi, 2010), pp. 1501–1527. MR 2827853
57. D. Gaboriau, What is ... cost? Notices Am. Math. Soc. **57**(10), 1295–1296 (2010). MR 2761803
58. D. Gaboriau, On the top-dimensional ℓ-Betti numbers. https://arxiv.org/abs/1909.01633
59. Ł. Grabowski, On Turing dynamical systems and the Atiyah problem. Invent. Math. **198**(1), 27–69 (2014). MR 3260857
60. Ł. Grabowski, Group ring elements with large spectral density. Math. Ann. **363**(1–2), 637–656 (2015). MR 3394391
61. R.I. Grigorčuk, On Burnside's problem on periodic groups. Funktsional. Anal. i Prilozhen. **14**(1), 53–54 (1980) (Russian). MR 565099
62. M. Gromov, Volume and bounded cohomology. Inst. Hautes Études Sci. Publ. Math. **56**, 5–99 (1982). MR 686042
63. M. Gromov, Large Riemannian manifolds, in *Curvature and Topology of Riemannian Manifolds* (Katata, 1985). Lecture Notes in Mathematics, vol. 1201 (Springer, Berlin, 1986), pp. 108–121. MR 859578
64. M. Gromov, Asymptotic invariants of infinite groups, in *Geometric Group Theory*, vol. 2 (Sussex, 1991). London Mathematical Society. Lecture Notes Series, vol. 182 (Cambridge University Press, Cambridge, 1993), pp. 1–295. MR 1253544
65. M. Gromov, Endomorphisms of symbolic algebraic varieties. J. Eur. Math. Soc. **1**(2), 109–197 (1999). MR 1694588

66. M. Gromov, *Metric Structures for Riemannian and Non-Riemannian Spaces*. Progress in Mathematics, vol. 152 (Birkhäuser, Boston, 1999). Based on the 1981 French original [MR0682063 (85e:53051)]; With appendices by M. Katz, P. Pansu, S. Semmes; Translated from the French by Sean Michael Bates. MR 1699320
67. M. Hall Jr., *The Theory of Groups* (Macmillan, New York, 1959). MR 0103215
68. A. Hatcher, *Algebraic Topology* (Cambridge University Press, Cambridge, 2002). MR 1867354
69. J. Hempel, *3-Manifolds*. Annals of Mathematics Studies, No. 86 (Princeton University Press/University of Tokyo Press, Princeton/Tokyo, 1976). MR 0415619
70. J. Hempel, Residual finiteness for 3-manifolds, in *Combinatorial Group Theory and Topology* (Alta, UT, 1984). Annals of Mathematics Studies, vol. 111 (Princeton University Press, Princeton, 1987), pp. 379–396. MR 895623
71. J. Hempel, Some 3-manifold groups with the same finite quotients (2014). arXiv:1409.3509
72. G. Herrmann, The L^2-Alexander torsion for Seifert fiber spaces. Arch. Math. (Basel) **109**(3), 273–283 (2017). MR 3687871
73. E. Hess, T. Schick, L^2-torsion of hyperbolic manifolds. Manuscripta Math. **97**(3), 329–334 (1998). MR 1654784
74. G. Higman, A finitely generated infinite simple group. J. Lond. Math. Soc. **26**, 61–64 (1951). MR 0038348
75. K.A. Hirsch, On infinite soluble groups. IV. J. Lond. Math. Soc. **27**, 81–85 (1952). MR 0044526
76. F. Hirzebruch, Automorphe Formen und der Satz von Riemann-Roch, in *Symposium Internacional de Topología Algebraica International Symposium on Algebraic Topology* (Universidad Nacional Autónoma de México and UNESCO, Mexico City, 1958), pp. 129–144 (German). MR 0103280
77. T. Hutchcroft, P. Gabor, Kazhdan groups have cost 1 (2018). arXiv:1810.11015
78. S. Illman, The equivariant triangulation theorem for actions of compact Lie groups. Math. Ann. **262**(4), 487–501 (1983). MR 696520
79. A. Jaikin-Zapirain, Approximation by subgroups of finite index and the Hanna Neumann conjecture. Duke Math. J. **166**(10), 1955–1987 (2017). MR 3679885
80. A. Jaikin-Zapirain, ℓ^2-Betti numbers and their analogues in positive characteristic, in *Proceedings of St. Andrews* (2017). http://verso.mat.uam.es/~andrei.jaikin/preprints/survey l2.pdf
81. A. Jaikin-Zapirain, The base change in the Atiyah and the Lück approximation conjectures (2018). http://verso.mat.uam.es/~andrei.jaikin/preprints/sac.pdf
82. A. Jaikin-Zapirain, D. López-Álvarez, The strong Atiyah conjecture for one-relator groups (2018). arXiv:1810.12135
83. V.F.R. Jones, *Von Neumann Algebras, 2009 to 2015*. Lecture Notes. https://math.vanderbilt.edu/jonesvf/VONNEUMANNALGEBRAS2015/VonNeumann2015.pdf
84. H. Kammeyer, L^2-invariants of nonuniform lattices in semisimple Lie groups. Algebr. Geom. Topol. **14**(4), 2475–2509 (2014). MR 3331619
85. H. Kammeyer, *Algebraic Topology I*. Lecture Notes (2016). http://www.math.kit.edu/iag7/~kammeyer/
86. H. Kammeyer, *Notes on the Abért–Nikolov Theorem on Rank Gradient and Cost* (2016). Blog post. https://perso.math.univ-toulouse.fr/jraimbau/2016/11/28/
87. H. Kammeyer, The shrinkage type of knots. Bull. Lond. Math. Soc. **49**(3), 428–442 (2017)
88. H. Kammeyer, A remark on torsion growth in homology and volume of 3-manifolds (2018). arXiv:1802.09244
89. H. Kammeyer, Profinite commensurability of S-arithmetic groups (2018). arXiv:1802.08559
90. H. Kammeyer, R. Sauer, S-arithmetic spinor groups with the same finite quotients and distinct ℓ^2-cohomology (2018). arXiv:1804.10604
91. H. Kammeyer, S. Kionke, J. Raimbault, R. Sauer, Profinite invariants of arithmetic groups (2019). arXiv:1901.01227

92. A. Kar, N. Nikolov, Rank gradient and cost of Artin groups and their relatives. Groups Geom. Dyn. **8**(4), 1195–1205 (2014). MR 3314944

93. A.S. Kechris, *Classical Descriptive Set Theory*. Graduate Texts in Mathematics, vol. 156 (Springer, New York, 1995). MR 1321597

94. S. Kionke, On p-adic limits of topological invariants (2008). Preprint. arXiv:1811.00356

95. S. Kionke, Lefschetz numbers of symplectic involutions on arithmetic groups. Pacific J. Math. **271**(2), 369–414 (2014). MR 3267534

96. S. Kionke, The growth of Betti numbers and approximation theorems (2017). arXiv:1709.00769

97. S. Kionke, Characters, L^2-Betti numbers and an equivariant approximation theorem. Math. Ann. **371**(1–2), 405–444 (2018). MR 3788853

98. M. Kreck, W. Lück, Topological rigidity for non-aspherical manifolds. Pure Appl. Math. Q. **5**(3), 873–914 (2009). Special Issue: In honor of Friedrich Hirzebruch. MR 2532709

99. D. Kyed, H.D. Petersen, S. Vaes, L^2-Betti numbers of locally compact groups and their cross section equivalence relations. Trans. Am. Math. Soc. **367**(7), 4917–4956 (2015). MR 3335405

100. M. Lackenby, Expanders, rank and graphs of groups. Israel J. Math. **146**, 357–370 (2005). MR 2151608

101. G. Levitt, On the cost of generating an equivalence relation. Ergodic Theory Dyn. Syst. **15**(6), 1173–1181 (1995). MR 1366313

102. T. Li, Rank and genus of 3-manifolds. J. Am. Math. Soc. **26**(3), 777–829 (2013). MR 3037787

103. H. Li, A. Thom, Entropy, determinants, and L^2-torsion. J. Am. Math. Soc. **27**(1), 239–292 (2014). MR 3110799

104. W. Li, W. Zhang, An L^2-Alexander invariant for knots. Commun. Contemp. Math. **8**(2), 167–187 (2006). MR 2219611

105. P.A. Linnell, Division rings and group von Neumann algebras. Forum Math. **5**(6), 561–576 (1993). MR 1242889

106. P. Linnell, T. Schick, Finite group extensions and the Atiyah conjecture. J. Am. Math. Soc. **20**(4), 1003–1051 (2007). MR 2328714

107. P. Linnell, W. Lück, R. Sauer, The limit of $\mathbf{F}_p$-Betti numbers of a tower of finite covers with amenable fundamental groups. Proc. Am. Math. Soc. **139**(2), 421–434 (2011). MR 2736326

108. Y. Liu, Degree of L^2-Alexander torsion for 3-manifolds. Invent. Math. **207**(3), 981–1030 (2017). MR 3608287

109. C. Löh, Simplicial volume. Bull. Manifold Atlas (2011). http://www.boma.mpim-bonn.mpg.de/

110. J. Lott, Heat kernels on covering spaces and topological invariants. J. Differ. Geom. **35**(2), 471–510 (1992). MR 1158345

111. J. Lott, The zero-in-the-spectrum question. Enseign. Math. (2) **42**(3–4), 341–376 (1996). MR 1426443

112. J. Lott, Deficiencies of lattice subgroups of Lie groups. Bull. Lond. Math. Soc. **31**(2), 191–195 (1999). MR 1664196

113. J. Lott, W. Lück, L^2-topological invariants of 3-manifolds. Invent. Math. **120**(1), 15–60 (1995). MR 1323981

114. W. Lück, L^2-Betti numbers of mapping tori and groups. Topology **33**(2), 203–214 (1994). MR 1273782

115. W. Lück, Approximating L^2-invariants by their finite-dimensional analogues. Geom. Funct. Anal. **4**(4), 455–481 (1994). MR 1280122

116. W. Lück, Dimension theory of arbitrary modules over finite von Neumann algebras and L^2-Betti numbers. I. Foundations. J. Reine Angew. Math. **495**, 135–162 (1998). MR 1603853

117. W. Lück, L^2-*Invariants: Theory and Applications to Geometry and K-Theory*. Ergebnisse der Mathematik und ihrer Grenzgebiete. 3. Folge, vol. 44 (Springer, Berlin, 2002). MR 1926649

118. W. Lück, Survey on classifying spaces for families of subgroups, in *Infinite Groups: Geometric, Combinatorial and Dynamical Aspects*. Progress in Mathematics, vol. 248 (Birkhäuser, Basel, 2005), pp. 269–322. MR 2195456

119. W. Lück, Twisting L^2-invariants with finite-dimensional representation (2015). arXiv:1510.00057

120. W. Lück, Approximating L^2-invariants by their classical counterparts. EMS Surv. Math. Sci. **3**(2), 269–344 (2016). MR 3576534

121. W. Lück, D. Osin, Approximating the first L^2-Betti number of residually finite groups. J. Topol. Anal. **3**(2), 153–160 (2011). MR 2819192

122. W. Lück, T. Schick, L^2-torsion of hyperbolic manifolds of finite volume. Geom. Funct. Anal. **9**(3), 518–567 (1999). MR 1708444

123. W. Lück, R. Sauer, C. Wegner, L^2-torsion, the measure-theoretic determinant conjecture, and uniform measure equivalence. J. Topol. Anal. **2**(2), 145–171 (2010). MR 2652905

124. R.C. Lyndon, P.E. Schupp, *Combinatorial Group Theory*. Classics in Mathematics (Springer, Berlin, 2001). Reprint of the 1977 edition. MR 1812024

125. G.A. Margulis, *Discrete Subgroups of Semisimple Lie Groups*. Ergebnisse der Mathematik und ihrer Grenzgebiete (3) [Results in Mathematics and Related Areas (3)], vol. 17 (Springer, Berlin, 1991). MR 1090825

126. S. Marshall, W. Müller, On the torsion in the cohomology of arithmetic hyperbolic 3-manifolds. Duke Math. J. **162**(5), 863–888 (2013). MR 3047468

127. S. Meskin, Nonresidually finite one-relator groups. Trans. Am Math. Soc. **164**, 105–114 (1972). MR 0285589

128. G.D. Mostow, *Strong Rigidity of Locally Symmetric Spaces*. Annals of Mathematics Studies, No. 78 (Princeton University Press/University of Tokyo Press, Princeton/Tokyo, 1973). MR 0385004

129. W. Müller, Analytic torsion and R-torsion for unimodular representations. J. Am. Math. Soc. **6**(3), 721–753 (1993). MR 1189689

130. W. Müller, J. Pfaff, Analytic torsion of complete hyperbolic manifolds of finite volume. J. Funct. Anal. **263**(9), 2615–2675 (2012). MR 2967302

131. W. Müller, J. Pfaff, Analytic torsion and L^2-torsion of compact locally symmetric manifolds. J. Differ. Geom. **95**(1), 71–119 (2013). MR 3128980

132. W. Müller, J. Pfaff, On the asymptotics of the Ray-Singer analytic torsion for compact hyperbolic manifolds. Int. Math. Res. Not. **13**, 2945–2983 (2013). MR 3072997

133. W. Müller, J. Pfaff, On the growth of torsion in the cohomology of arithmetic groups. Math. Ann. **359**(1–2), 537–555 (2014). MR 3201905

134. W. Müller, J. Pfaff, The analytic torsion and its asymptotic behaviour for sequences of hyperbolic manifolds of finite volume. J. Funct. Anal. **267**(8), 2731–2786 (2014). MR 3255473

135. B. Nica, Linear groups—Malcev's theorem and Selberg's lemma (2013). arXiv:1306.2385

136. N. Nikolov, Algebraic properties of profinite groups (2011). arXiv:1108.5130

137. N. Nikolov, D. Segal, On finitely generated profinite groups. I. Strong completeness and uniform bounds. Ann. Math. (2) **165**(1), 171–238 (2007). MR 2276769

138. nLab authors, *von Neumann Algebra* (2019). http://ncatlab.org/nlab/show/von+Neumann+algebra

139. M. Olbrich, L^2-invariants of locally symmetric spaces. Doc. Math. **7**, 219–237 (2002). MR 1938121

140. D.S. Ornstein, B. Weiss, Ergodic theory of amenable group actions. I. The Rohlin lemma. Bull. Am. Math. Soc. (N.S.) **2**(1), 161–164 (1980). MR 551753

141. D. Pape, A short proof of the approximation conjecture for amenable groups. J. Funct. Anal. **255**(5), 1102–1106 (2008). MR 2455493

142. H.D. Petersen, L2-Betti numbers of locally compact groups, PhD Thesis, Department of Mathematical Sciences, Faculty of Science, University of Copenhagen, 2012. http://www.math.ku.dk/noter/filer/phd13hdp.pdf

143. H.D. Petersen, A. Valette, L^2-Betti numbers and Plancherel measure. J. Funct. Anal. **266**(5), 3156–3169 (2014). MR 3158720

144. M. Pichot, Semi-continuity of the first l^2-Betti number on the space of finitely generated groups. Comment. Math. Helv. **81**(3), 643–652 (2006). MR 2250857

145. M. Pichot, T. Schick, A. Zuk, Closed manifolds with transcendental L^2-Betti numbers. J. Lond. Math. Soc. (2) **92**(2), 371–392 (2015). MR 3404029
146. V. Platonov, A. Rapinchuk, *Algebraic Groups and Number Theory*. Pure and Applied Mathematics, vol. 139 (Academic, Boston, 1994). Translated from the 1991 Russian original by Rachel Rowen. MR 1278263
147. G. Prasad, A.S. Rapinchuk, Developments on the congruence subgroup problem after the work of Bass, Milnor and Serre (2008). arXiv:0809.1622
148. P. Przytycki, D.T. Wise, Mixed 3-manifolds are virtually special. J. Am. Math. Soc. **31**(2), 319–347 (2018). MR 3758147
149. A.A. Ranicki, *Notes on Reidemeister Torsion*. Department of Mathematics and Statistics University of Edinburgh. http://www.maths.ed.ac.uk/~aar/papers/torsion.pdf
150. M. Reed, B. Simon, *Methods of Modern Mathematical Physics. I. Functional Analysis*, 2nd edn. (Academic, New York, 1980). MR 751959
151. A.W. Reid, Profinite properties of discrete groups, in *Groups St. Andrews 2013*. London Mathematics Society. Lecture Note Series, vol. 422 (Cambridge University Press, Cambridge, 2015), pp. 73–104. MR 3445488
152. L. Ribes, P. Zalesskii, *Profinite Groups*. Ergebnisse der Mathematik und ihrer Grenzgebiete. 3. Folge. A Series of Modern Surveys in Mathematics, vol. 40 (Springer, Berlin, 2000). MR 1775104
153. W. Rudin, *Real and Complex Analysis*, 3rd edn. (McGraw-Hill, New York, 1987). MR 924157
154. S. Sakai, A characterization of W^*-algebras. Pacific J. Math. **6**, 763–773 (1956). MR 0084115
155. R. Sauer, Amenable covers, volume and L^2-Betti numbers of aspherical manifolds. J. Reine Angew. Math. **636**, 47–92 (2009). MR 2572246
156. T. Schick, Integrality of L^2-Betti numbers. Math. Ann. **317**(4), 727–75 (2000). MR 1777117
157. T. Schick, L^2-determinant class and approximation of L^2-Betti numbers. Trans. Am. Math. Soc. **353**(8), 3247–3265 (2001). MR 1828605
158. T. Schick, Erratum: "Integrality of L^2-Betti numbers". Math. Ann. **322**(2), 421–422 (2002). MR 1894160
159. K. Schmidt, *Dynamical Systems of Algebraic Origin*. Progress in Mathematics, vol. 128 (Birkhäuser, Basel, 1995). MR 1345152
160. P. Scholze, On torsion in the cohomology of locally symmetric varieties. Ann. Math. (2) **182**(3), 945–1066 (2015). MR 3418533
161. K. Schreve, The strong Atiyah conjecture for virtually cocompact special groups. Math. Ann. **359**(3–4), 629–636 (2014). MR 3231009
162. M.H. Şengün, On the integral cohomology of Bianchi groups. Exp. Math. **20**(4), 487–505 (2011). MR 2859903
163. M.H. Şengün, On the torsion homology of non-arithmetic hyperbolic tetrahedral groups. Int. J. Number Theory **8**(2), 311–320 (2012). MR 2890481
164. B. Sury, T.N. Venkataramana, Generators for all principal congruence subgroups of SL($n\mathbf{Z}$) with $n \geq 3$. Proc. Am. Math. Soc. **122**(2), 355–358 (1994). MR 1239806
165. W. Thurston, *The Geometry and Topology of 3-Manifolds*. Lecture Notes (1980). http://library.msri.org/books/gt3m/
166. W. Thurston, A norm for the homology of 3-manifolds. Mem. Am. Math. Soc. **59**(339), i–vi and 99–130 (1986). MR 823443
167. T. tom Dieck, *Algebraic Topology*. EMS Textbooks in Mathematics (European Mathematical Society, Zürich, 2008). MR 2456045
168. K. Vogtmann, Automorphisms of free groups and outer space, in *Proceedings of the Conference on Geometric and Combinatorial Group Theory, Part I* (Haifa, 2000), 2002, pp. 1–31. MR 1950871
169. C.T.C. Wall, *Surgery on Compact Manifolds*, 2nd edn. Mathematical Surveys and Monographs, vol. 69 (American Mathematical Society, Providence, 1999). Edited and with a foreword by A.A. Ranicki. MR 1687388

170. H.C. Wang, Topics on totally discontinuous groups, in *Symmetric Spaces* (Short Courses, Washington University, St. Louis, 1969–1970). Pure and Applied Mathematics, vol. 8 (Dekker, New York, 1972), pp. 459–487. MR 0414787

171. B. Weiss, Sofic groups and dynamical systems. Sankhyā Ser. A **62**(3), 350–359 (2000). Ergodic theory and harmonic analysis (Mumbai, 1999). MR 1803462

172. G. Wilkes, Profinite rigidity for Seifert fibre spaces. Geom. Dedicata **188**, 141–163 (2017). MR 3639628

173. D.T. Wise, Research announcement: the structure of groups with a quasiconvex hierarchy. Electron. Res. Announc. Math. Sci. **16**, 44–55 (2009). MR 2558631

174. D.T. Wise, The structure of groups with a quasiconvex hierarchy (2012). https://drive.google.com/file/d/0B45cNx80t5-2T0twUDFxVXRnQnc/view

175. D. Witte Morris, *Introduction to Arithmetic Groups* (Deductive Press, Public Domain, 2015). MR 3307755

List of Notation

$(\Gamma_i \backslash X)_{<R}$	R-thin part of $\Gamma_i \backslash X$, p. 107
$\chi(X)$	Euler characteristic, p. 4
$\chi^{(2)}(X)$	ℓ^2-Euler characteristic, p. 50
$\chi_-(\Sigma)$	complexity of the surface Σ, p. 140
χ_A	characteristic function of the set A, p. 20
$\mathrm{cost}(E)$	cost of the equivalence relation E, p. 110
$\mathrm{cost}(G)$	cost of the group G, p. 110
$\mathrm{cost}(\Omega \curvearrowleft G)$	cost of the measurable group action $\Omega \curvearrowleft G$, p. 110
$\mathrm{def}(G)$	deficiency of a finitely presented group G, p. 81
$\deg \tau^{(2)}(N, \phi)$	asymptotic degree of $\tau^{(2)}(N, \phi)$, p. 142
$\deg_S(x)$	number of outgoing edges from x in the graph S, p. 110
$\delta(G)$	deficiency of the semisimple Lie group G, p. 136
δ_λ	Dirac measure supported on $\{\lambda\}$, p. 93
$\det_{\mathcal{R}(G)} T$	Fuglede–Kadison determinant of T, p. 118
$\dim_{kG}^{\mathrm{Ore}}$	Ore dimension over the group ring kG, p. 106
$\dim_{\mathcal{L}(G)}$	von Neumann dimension of right Hilbert module, p. 53
$\dim_{\mathcal{R}(G)}$	von Neumann dimension of left Hilbert module, p. 29
$\ell^2 G$	Hilbert space of square integrable sums over G, p. 3
ℓ^2	square summable sequences of complex numbers, p. 11
$\ell^\infty G$	Banach space of bounded functions on G, p. 107
$\mathrm{gcost}(\Omega \curvearrowleft G)$	groupoid cost of the measurable action $\Omega \curvearrowleft G$, p. 113
$\mathrm{ind}_{G_0}^G$	induction functor, p. 52
λ, ρ	left and right regular representations, p. 19
$\langle \cdot, \cdot \rangle$	inner product, p. 9
$\mathbb{C}G$	group algebra with complex coefficients, p. 19
$\mathbb{RP}^\infty$	infinite dimensional projective space, p. 78
$\mathbb{Z}(G/H_j)^{H_i}$	H_i-invariants of the $\mathbb{Z}G$-module $\mathbb{Z}(G/H_j)$, p. 43
$\mathbb{H}^n$	n-dimensional hyperbolic space, p. 3
$\mathbb{P}$	polytope homomorphism, p. 143

© Springer Nature Switzerland AG 2019
H. Kammeyer, *Introduction to ℓ^2-invariants*, Lecture Notes in Mathematics 2247,
https://doi.org/10.1007/978-3-030-28297-4

$\mathbb{T}^k$	k-dimensional torus,　p. 4
G	linear algebraic group,　p. 151
PM, **T**M	projective/torsion part of the $\mathcal{R}(G)$-module M,　p. 74
Spin(q)	spinor group of the quadratic form q,　p. 161
T^n	n-dimensional algebraic torus,　p. 152
$\mathcal{ALL}$	family of all subgroups,　p. 67
$\mathcal{AME}$	family of all amenable subgroups,　p. 77
$\mathcal{C}$	Linnell's class of groups,　p. 60
$\mathcal{D}$	Schick's class of groups $\mathcal{D}$,　p. 125
$\mathcal{E}$	class of elementary amenable groups,　p. 60
$\mathcal{F}$	a general family of subgroups,　p. 67
$\mathcal{FIN}$	family of all finite subgroups,　p. 67
$\mathcal{L}(G)$-mod	category of left Hilbert modules,　p. 53
$\mathcal{L}^2[a, b]$	square integrable functions on the interval $[a, b]$,　p. 11
$\mathcal{L}H$	right Hilbert module made left Hilbert module,　p. 53
$\mathcal{O}(U)$	algebra of holomorphic functions on $U \subseteq \mathbb{C}$,　p. 90
$\mathcal{O}_F$	ring of integers in number field F,　p. 124
$\mathcal{P}_{\mathbb{Z}}^{\mathrm{Wh}}(H_1(G)_{\mathrm{free}})$	integral polytope group in $H_1(G)_{\mathrm{free}}$,　p. 143
$\mathcal{R}(G)$, $\mathcal{L}(G)$	right and left group von Neumann algebra of G,　p. 19
$\mathcal{TRIV}$	trivial family of subgroups,　p. 67
$\mathcal{VCYC}$	family of all virtually cyclic subgroups,　p. 67
$\mathfrak{g}$	Lie algebra of the Lie group G,　p. 136
$\mathfrak{k}$	Lie algebra of the Lie group K,　p. 136
$\mathrm{tr}_{\mathcal{R}}(G)$	von Neumann trace,　p. 24
$\mathcal{B}(\sigma(T), \mathbb{C})$	bounded $\mathbb{C}$-valued Borel functions on $\sigma(T)$,　p. 92
$\min \mathrm{vol}(M)$	minimal volume of the smooth manifold M,　p. 116
μ_T	basic measure of the bounded operator T,　p. 93
$\mu_{x,T}$	spectral measure of T associated with x,　p. 92
$\mathrm{Out}(F_n)$	group of outer automorphisms of F_n,　p. 109
$\overline{\mathrm{im}\, T}$	closure of image of operator T,　p. 17
$\overline{X}_N$	Galois covering of X associated with $N \trianglelefteq \pi_1 X$,　p. 1
∂T	boundary of tree T,　p. 103
rank_k	rank of a semisimple algebraic group,　p. 152
$\mathrm{res}_{G_0}^G$	restriction functor,　p. 27
$\rho(C_*)$	Reidemeister torsion of the chain complex C_*,　p. 129
$\rho(X; V)$	Reidemeister torsion of X with coefficients in V,　p. 131
$\rho^M(\Gamma_i \backslash X)$	integral torsion of $\Gamma_i \backslash X$ with coefficients in M,　p. 153
$\rho^{(2)}(\varphi)$	ℓ^2-torsion of the group automorphism φ,　p. 137
$\rho^{(2)}(C_*^{(2)})$	ℓ^2-torsion of the chain complex $C_*^{(2)}$,　p. 132
$\rho^{(2)}(G)$	ℓ^2-torsion of the group G,　p. 135
$\rho^{(2)}(X)$, $\rho^{(2)}(G \curvearrowright X)$	ℓ^2-torsion of the G-CW complex X,　p. 133
$\rho^{\mathbb{Z}}(X)$	integral torsion of the CW complex X,　p. 145
$\rho_u(X)$	universal ℓ^2-torsion of the G-CW complex X,　p. 143
$\mathrm{sh}(G_i g)$	shadow of the right coset $G_i g$,　p. 103

$\sigma(T)$	spectrum of the bounded operator T, p. 89
$\tau^{(2)}(N, \gamma, \phi)$	ℓ^2-Alexander torsion, p. 139
$\tau^{(2)}(N, \phi)$	full ℓ^2-Alexander torsion of N w.r.t. ϕ, p. 138
$\tau_{\text{weak}}, \tau_{\text{strong}}, \tau_{\text{norm}}$	weak, strong, and norm topology on $B(H)$, p. 22
$\text{lcm}(G)$	least common multiple of finite subgroup orders, p. 59
mod-$\mathcal{R}(G)$	category of right Hilbert modules, p. 53
pr_U	orthogonal projection onto the subspace U, p. 25
$\text{tr}_{\mathcal{L}(G)}$	trace in left group von Neumann algebra, p. 53
EG	classifying space of G for proper actions, p. 70
Δ_K	Alexander polynomial of the knot K, p. 139
$\Delta_n^{(2)}$	n-th ℓ^2-Laplacian, p. 55
$\varrho(T)$	resolvent set of the bounded operator T, p. 89
Σ_g	closed surface of genus g, p. 56
$\Sigma_{g,d}$	surface of genus g with d punctures, p. 56
$\|T\|$	operator norm, p. 16
$\text{vol}(\Gamma \backslash X)$	volume of the locally symmetric space $\Gamma \backslash X$, p. 152
$\text{Wh}^w(G)$	weak Whitehead group of G, p. 143
$\widehat{G}$	profinite completion of the group G, p. 156
$\widetilde{B}^{(2)}$	subset of $B^{(2)}$ for simply connected spaces, p. 57
$\widetilde{X}$	universal covering of X, p. 1
$B(\ell^2 G)^\lambda, B(\ell^2 G)^\lambda$	left and right equivariant bounded operators, p. 19
$B(H)$	bounded operators on H, p. 16
$B(H, K)$	bounded operators from H to K, p. 16
$B(m, n)$	Baumslag–Solitar group, p. 88
$B^{(2)}$	set of possible values of ℓ^2-Betti numbers, p. 57
$b_n^{(2)}(G)$	n-th ℓ^2-Betti number of the group G, p. 77
$b_n^{(2)}(G, \mathcal{F})$	n-th ℓ^2-Betti number of G w.r.t. the family $\mathcal{F}$, p. 71
$b_n^{(2)}(X), b_n^{(2)}(G \curvearrowright X)$	n-th ℓ^2-Betti number of the G-CW complex X, p. 46
$b_n^{(2)}(X, A)$	relative n-th ℓ^2-Betti number of the pair (X, A), p. 56
$B_R^{(2)}(G)$	set of kernel dimensions for RG-matrices, p. 59
$b_{(2)}^n(X)$	n-th cohomological ℓ^2-Betti number of X, p. 55
BG	classifying space (quotient), p. 70
$C[a, b]$	continuous functions on the interval $[a, b]$, p. 10
$C_{(2)}^*(X)$	ℓ^2-cochain complex of the G-CW complex X, p. 53
$C_p^1[-\pi, \pi]$	continuously differentiable periodic function, p. 14
$C_*^{(2)}(X)$	ℓ^2-chain complex of the G-CW complex X, p. 44
$C_*^{\text{sing}}(X)$	singular chain complex of the $(G\text{-})$space X, p. 75
$C_*(X)$	cellular chain complex of a $(G\text{-})$CW complex, p. 38
$C_{\text{odd/even}}$	odd/even part of the chain complex C_*, p. 129
$d(G)$	rank of a group, p. 108
D_∞	infinite dihedral group, p. 38
$e(S)$	edge measure of the graph S, p. 110
$E_x, E_{x,y}$	evaluation maps, p. 16

Index

© Springer Nature Switzerland AG 2019
H. Kammeyer, *Introduction to ℓ^2-invariants*, Lecture Notes in Mathematics 2247,
https://doi.org/10.1007/978-3-030-28297-4

LECTURE NOTES IN MATHEMATICS   🐎 Springer

Editors in Chief: J.-M. Morel, B. Teissier;

Editorial Policy

1. Lecture Notes aim to report new developments in all areas of mathematics and their applications – quickly, informally and at a high level. Mathematical texts analysing new developments in modelling and numerical simulation are welcome.

   Manuscripts should be reasonably self-contained and rounded off. Thus they may, and often will, present not only results of the author but also related work by other people. They may be based on specialised lecture courses. Furthermore, the manuscripts should provide sufficient motivation, examples and applications. This clearly distinguishes Lecture Notes from journal articles or technical reports which normally are very concise. Articles intended for a journal but too long to be accepted by most journals, usually do not have this "lecture notes" character. For similar reasons it is unusual for doctoral theses to be accepted for the Lecture Notes series, though habilitation theses may be appropriate.

2. Besides monographs, multi-author manuscripts resulting from SUMMER SCHOOLS or similar INTENSIVE COURSES are welcome, provided their objective was held to present an active mathematical topic to an audience at the beginning or intermediate graduate level (a list of participants should be provided).

   The resulting manuscript should not be just a collection of course notes, but should require advance planning and coordination among the main lecturers. The subject matter should dictate the structure of the book. This structure should be motivated and explained in a scientific introduction, and the notation, references, index and formulation of results should be, if possible, unified by the editors. Each contribution should have an abstract and an introduction referring to the other contributions. In other words, more preparatory work must go into a multi-authored volume than simply assembling a disparate collection of papers, communicated at the event.

3. Manuscripts should be submitted either online at www.editorialmanager.com/lnm to Springer's mathematics editorial in Heidelberg, or electronically to one of the series editors. Authors should be aware that incomplete or insufficiently close-to-final manuscripts almost always result in longer refereeing times and nevertheless unclear referees' recommendations, making further refereeing of a final draft necessary. The strict minimum amount of material that will be considered should include a detailed outline describing the planned contents of each chapter, a bibliography and several sample chapters. Parallel submission of a manuscript to another publisher while under consideration for LNM is not acceptable and can lead to rejection.

4. In general, **monographs** will be sent out to at least 2 external referees for evaluation.

   A final decision to publish can be made only on the basis of the complete manuscript, however a refereeing process leading to a preliminary decision can be based on a pre-final or incomplete manuscript.

   Volume Editors of **multi-author works** are expected to arrange for the refereeing, to the usual scientific standards, of the individual contributions. If the resulting reports can be forwarded to the LNM Editorial Board, this is very helpful. If no reports are forwarded or if other questions remain unclear in respect of homogeneity etc, the series editors may wish to consult external referees for an overall evaluation of the volume.

5. Manuscripts should in general be submitted in English. Final manuscripts should contain at least 100 pages of mathematical text and should always include

   – a table of contents;
   – an informative introduction, with adequate motivation and perhaps some historical remarks: it should be accessible to a reader not intimately familiar with the topic treated;
   – a subject index: as a rule this is genuinely helpful for the reader.
   – For evaluation purposes, manuscripts should be submitted as pdf files.

6. Careful preparation of the manuscripts will help keep production time short besides ensuring satisfactory appearance of the finished book in print and online. After acceptance of the manuscript authors will be asked to prepare the final LaTeX source files (see LaTeX templates online: https://www.springer.com/gb/authors-editors/book-authors-editors/manuscriptpreparation/5636) plus the corresponding pdf- or zipped ps-file. The LaTeX source files are essential for producing the full-text online version of the book, see http://link.springer.com/bookseries/304 for the existing online volumes of LNM). The technical production of a Lecture Notes volume takes approximately 12 weeks. Additional instructions, if necessary, are available on request from lnm@springer.com.

7. Authors receive a total of 30 free copies of their volume and free access to their book on SpringerLink, but no royalties. They are entitled to a discount of 33.3 % on the price of Springer books purchased for their personal use, if ordering directly from Springer.

8. Commitment to publish is made by a *Publishing Agreement*; contributing authors of multiauthor books are requested to sign a *Consent to Publish form*. Springer-Verlag registers the copyright for each volume. Authors are free to reuse material contained in their LNM volumes in later publications: a brief written (or e-mail) request for formal permission is sufficient.

Addresses:
Professor Jean-Michel Morel, CMLA, École Normale Supérieure de Cachan, France
E-mail: moreljeanmichel@gmail.com

Professor Bernard Teissier, Equipe Géométrie et Dynamique,
Institut de Mathématiques de Jussieu – Paris Rive Gauche, Paris, France
E-mail: bernard.teissier@imj-prg.fr

Springer: Ute McCrory, Mathematics, Heidelberg, Germany,
E-mail: lnm@springer.com

Printed in the United States
By Bookmasters